Scale-Up and Automation in Plant Propagation

Scale-Up and Automation in Plant Propagation

Cell Culture and Somatic Cell Genetics of Plants, Volume 8

Edited by

INDRA K. VASIL
Laboratory of Plant Cell and Molecular Biology
Department of Vegetable Crops
University of Florida
Gainesville, Florida

ACADEMIC PRESS, INC.
Harcourt Brace Jovanovich, Publishers
San Diego New York Boston London Sydney Tokyo Toronto

Front cover illustration: An adaptation of Figure 6, Chapter 8, a staged reactor design in a spin-filter system. (For details see p. 174.)

Academic Press, Inc.
San Diego, California 92101

United Kingdom Edition published by
Academic Press Limited
24–28 Oval Road, London NW1 7DX

Library of Congress Cataloging-in-Publication Data
(Revised for volume 8)

Cell culture and somatic cell genetics of plants.

Vols. 4-5 edited by Friedrich Constabel, Indra K. Vasil; v. 6 edited by Jeff Schell and Indra K. Vasil.
Includes bibliographies and indexes.
Contents: v. 1. Laboratory procedures and their applications -- v. 2. Cell growth, nutrition, cytodifferentiation, and cryopreservation -- [etc.] -- v. 8. Scale-up and automation in plant propagation.
1. Plant cell culture. 2. Plant cytogenetics.
I. Vasil, I. K. II. Constabel, F. III. Schell, Jeff.
QK725.C37 1984 581'.07'24 83-21538
ISBN 0-12-433426-1

PRINTED IN THE UNITED STATES OF AMERICA
91 92 93 94 9 8 7 6 5 4 3 2 1

Contents

7 Problems and Perspectives for the Automated Propagation of Fruit Trees

ALESSIO MARTINELLI

8 Bioreactors and Image Analysis for Scale-Up and Plant Propagation

DANA CAZZULINO, HENRIK PEDERSEN, and CHEE-KOK CHIN

9 Automated Evaluation of Somatic Embryogenesis in Sweet Potato by Machine Vision

ROY C. HARRELL and DANIEL J. CANTLIFFE

10 Control of Contamination in Automated Plant Protopagation

ALAN C. CASSELLS

Contributors

Numbers in parentheses indicate the pages on which the authors' contributions begin.

HAROLD L. BREWER (245), Crop Systems Research Unit, U.S. Dept. of Agriculture, Coastal Plains Experiment Station, Tifton, Georgia 31793

DANIEL J. CANTLIFFE (179), Agricultural Engineering and Vegetable Crops Departments, University of Florida, Gainesville, Florida 32611

WILLIAM CARLSON (75), Technology Center, Weyerhaeuser Company, Tacoma, Washington 98477

CAROLYN CARPENTER (75), Technology Center, Weyerhaeuser Company, Tacoma, Washington 98477

ALAN C. CASSELLS (197), Department of Plant Science, University College, Cork, Ireland

DANA CAZZULINO (147), Department of Chemical and Biochemical Engineering and Department of Horticulture, Rutgers, The State University of New Jersey, New Brunswick, New Jersey 08102

CHEE-KOK CHIN (147), Department of Chemical and Biochemical Engineering and Department of Horticulture, Rutgers, The State University of New Jersey, New Brunswick, New Jersey 08102

IRWIN Y. E. CHU (7), Twyford International, Inc., Santa Paula, California 93060

JO ANN FUJII (36), Calgene, Inc., Davis, California 95616

NOBUYUKI FUJITA (231), Mechanism and Control Technology Department, Manufacturing Engineering Laboratory, Toshiba Corporation, Isogo-ku, Yokohama 235, Japan

PRAMOD K. GUPTA (75), Technology Center, Weyerhaeuser Company, Tacoma, Washington 98477

ROY C. HARRELL (179), Agricultural Engineering and Vegetable Crops Departments, University of Florida, Gainesville, Florida 32611

ROBERT D. HARTMAN (7), Twyford International, Inc., Santa Paula, California 93060
PETER J. JOYCE (95), Department of Horticulture, University of Wisconsin-Madison, Madison, Wisconsin 53706
ATSUSHI KINASE (231), Mechanism and Control Technology Department, Manufacturing Engineering Laboratory, Toshiba Corporation, Isogo-ku Yokohama 235, Japan
TOYOKI KOZAI (213), Laboratory of Horticultural Engineering, Department of Horticulture, Chiba University, Matsudo, Chiba 271, Japan
MARY KREITINGER (75), Technology Center, Weyerhaeuser Company, Tacoma, Washington 98477
SHARON L. KURTZ (7), Twyford International, Inc., Santa Paula, California 93060
ALESSIO MARTINELLI (133), Consorzio Italiano Vivaisti, Loc. Boattone - Stat. Romea Km 116, 44020 S. Giuseppe di Comacchio (Ferrara), Italy
BRENT H. MCCOWN (95), Department of Horticulture, University of Wisconsin-Madison, Madison, Wisconsin 53706
YOSHIYUKI MIWA (111), Department of Science and Engineering, Waseda University, Tokyo, Japan
HENRIK PEDERSEN (147), Department of Chemical and Biochemical Engineering and Department of Horticulture, Rutgers, The State University of New Jersey, New Brunswick, New Jersey 08102
GERALD PULLMAN (75), Technology Center, Weyerhaeuser Company, Tacoma, Washington 98477
KEITH REDENBAUGH (35), Calgene, Inc., Davis, California 95616
DAVID SLADE (35), Calgene, Inc., Davis, California 95616
BRAD SWEDLUND (111), Native Plants, Inc., Salt Lake City, Utah 84112
SHINSAKU TAKAYAMA (111), Department of Biological Science and Technology, Tokai University, 317 Nishino, Numazu, 410-03 Shizuoka, Japan
ROGER TIMMIS (75), Technology Center, Weyerhaeuser Company, Tacoma, Washington 98477
INDRA K. VASIL (1), Laboratory of Plant Cell and Molecular Biology, Department of Vegetable Crops, University of Florida, Gainesville, Florida 32611-0514
MIKE YANCEY (75), Technology Center, Weyerhaeuser Company, Tacoma, Washington 98477

General Preface

Recent advances in the techniques and applications of plant cell culture and plant molecular biology have created unprecedented opportunities for the genetic manipulation of plants. The potential impact of these novel and powerful biotechnologies on the genetic improvement of crop plants has generated considerable interest, enthusiasm, and optimism in the scientific community and is in part responsible for the rapidly expanding biotechnology industry.

The anticipated role of biotechnology in agriculture is based not on the actual production of any genetically superior plants, but on elegant demonstrations in model experimental systems that new hybrids, mutants, and genetically engineered plants can be obtained by these methods and the presumption that the same procedures can be adapted successfully for important crop plants. However, serious problems exist in the transfer of this technology to crop species.

Most of the current strategies for the application of biotechnology to crop improvement envisage the regeneration of whole plants from single, genetically altered cells. In many instances this requires that specific agriculturally important genes be identified and characterized, that they be cloned, that their regulatory and functional controls be understood, and that plants be regenerated from single cells in which such gene material has been introduced and integrated in a stable manner.

Knowledge of the structure, function, and regulation of plant genes is scarce, and basic research in this area is still limited. On the other hand, a considerable body of knowledge has accumulated in the last fifty years on the isolation and culture of plant cells and tissues. For example, it is possible to regenerate plants from tissue cultures of many plant species, including several important agricultural crops. These procedures are now widely used in large-scale rapid clonal propagation of plants. Plant cell culture techniques also allow the isolation of mutant cell lines and plants, the generation of somatic hybrids by protoplast fusion, and the regeneration of genetically engineered plants from single transformed cells.

Many national and international meetings have been the forums for discussion of the application of plant biotechnology to agriculture. Neither the basic techniques nor the biological principles of plant cell culture are generally included in these discussions or their published proceedings. Following the very enthusiastic reception accorded the two volumes entitled "Perspectives in Plant Cell and Tissue Culture" that were published as supplements to the *International Review of Cytology* in 1980, I was approached by Academic Press to consider the feasibility of publishing a treatise on plant cell culture. Because of the rapidly expanding interest in the subject both in academia and in industry, I was convinced that such a treatise was needed and would be useful. No comprehensive work of this nature is available or has been attempted previously.

The organization of the treatise is based on extensive discussions with colleagues, the advice of a distinguished editorial advisory board, and suggestions provided by anonymous reviewers to Academic Press. However, the responsibility for the final choice of subject matter included in the different volumes, and of inviting authors for various chapters, is mine. The basic premise on which this treatise is based is that knowledge of the principles of plant cell culture is critical to their potential use in biotechnology. Accordingly, descriptions and discussion of all aspects of modern plant cell culture techniques and research are included in the treatise. The first volume describes every major laboratory procedure used in plant cell culture and somatic cell genetics research, including variations of a single procedure adapted for important crop plants. The second and third volumes are devoted to the nutrition and growth of plant cell cultures and to the important subject of generating and recovering variability from cell cultures. An entirely new approach is used in the treatment of this subject by including not only spontaneous variability arising during culture, but also variability created by protoplast fusion, genetic transformation, etc. Future volumes are envisioned to cover most other relevant and current areas of research in plant cell culture and its uses in biotechnology.

In addition to the very comprehensive treatment of the subject, the uniqueness of these volumes lies in the fact that all the chapters are prepared by distinguished scientists who have played a major role in the development and/or uses of specific laboratory procedures and in key fundamental as well as applied studies of plant cell and tissue culture.

This allows a deep insight, as well as a broad perspective, based on personal experience. The volumes are designed as key reference works to provide extensive as well as intensive information on all aspects of plant cell and tissue culture not only to those newly entering the field but also to experienced researchers.

Indra K. Vasil

Preface

Regeneration of plants from cultured cells, tissues, and organs is a unique property of higher plants, and has recently become a critical component of plant biotechnology. Two earlier volumes of this treatise, Volumes 1 and 3, provided detailed accounts of plant regeneration from a wide variety of species. Large-scale propagation of plants from cultured shoot meristems, commonly described as micropropagation, has been the most widely used aspect of plant biotechnology for more than two decades.

The micropropagation industry has been only marginally profitable, and is still largely restricted to a few herbaceous ornamental species. No significant improvements in this technology, which still relies heavily on time-consuming and expensive manual labor, have taken place. On the other hand, rising labor, land, energy, and other expenses have increased the cost of micropropagation, without improving the quality of the product. As a result, the industry has failed to grow and has become static.

Growth and expansion of the industry are certainly possible, and can be achieved by reducing the cost of micropropagation to 25% or less of the present level, and opening it to include the vast market of transplant vegetables and fruit, spice, plantation, and forest species. Such a drastic reduction in the cost of micropropagated plants can be realized by automation of the various components of the micropropagation process, thereby reducing the overwhelming dependence on manual labor.

This volume summarizes and evaluates many innovative contemporary methods of automation and scale-up of plant propagation *in vitro*. It is my hope that the critical needs of the industry and future growth prospects will encourage further experimentation and that one or more of the procedures outlined in this volume—with modifications—will become a common feature of all large-scale micropropagation enterprises in the future.

I am grateful to Bob Levin, the inventor of the Vitromatic System of micropropagation, for sharing with me his infectious enthusiasm for the exciting and challenging field of automated plant propagation, and for his assistance in the organization of this volume. Thanks are also due to all the contributors for providing well-written manuscripts on schedule, thus lessening my editorial burden.

Indra K. Vasil

Cell Culture and Somatic Cell Genetics of Plants Contents of Previous Volumes

Laboratory Procedures and Their Applications, Volume 1

Cell Growth, Nutrition, Cytodifferentiation and Cryopreservation, Volume 2

Plant Regeneration and Genetic Variability, Volume 3

Part I Regeneration

Molecular Biology of Plant Nuclear Genes, Volume 6

The Molecular Biology of Plastids, Volume 7A

The Photosynthetic Apparatus: Molecular Biology and Operation, Volume 7B

Part I

Part II

CHAPTER **1**

Rationale for the Scale-Up and Automation of Plant Propagation

Indra K. Vasil

Laboratory of Plant Cell and Molecular Biology
Department of Vegetable Crops
University of Florida
Gainesville, Florida

Large scale clonal propagation of plants, popularly described as micropropagation, is the first major practical application of plant biotechnology. The beginnings of this industry go back about 30 years and are based on three important discoveries. The first critical finding was that cultured shoot meristems give rise to virus-free plants (Morel and Martin, 1952). This procedure, which was subsequently employed for virus elimination and the induction of protocorm formation in shoot tip cultures of orchids (Morel, 1960, 1965), is widely used for the clonal propagation of many orchids. Another milestone was the elucidation of the role of cytokinins in shoot morphogenesis (Skoog and Miller, 1957) and in the inhibition of apical dominance, which in turn released the axillary meristems from dormancy (Sachs and Thimann, 1964). The successful application of these principles to the *in vitro* multiplication of plants by micropropagation has been a key factor in the development of this technology into an important worldwide industry which produces 250–300 million units annually. The technology is simple, involving only the culture of isolated shoot buds (shoot meristems with several leaf primordia; shoot meristems with only one or two leaf primordia must be used in those instances in which virus elimination is desired) on nutrient media containing cytokinins to stimulate the development of multiple axillary buds by inhibiting apical dominance (see Chapter 2, this volume). Three- to sixfold increases in shoot number can be obtained every 4–6 weeks (less than a 3-fold increase is economically unacceptable), resulting in the production of one to several million plants from each cultured shoot meristem in one year. No elaborate facilities are needed or used in micropropagation. The simplicity of the methodology has allowed hundreds of small and a few large facilities to be developed in many countries throughout the world, including some very simple but productive village level operations in Southeast Asia.

There are two other methods of *in vitro* plant propagation. These are based on the *de novo* formation of shoot meristems in callus cultures, and on the induction of somatic embryogenesis (Vasil and Vasil, 1980). Neither of these procedures is currently used for large-scale production of plants. *De novo* formation of shoot meristems in callus tissues is generally avoided because of the high incidence of genetically variant plants and even chimeras. On the other hand, somatic embryogenesis is potentially the most efficient and economic method for large-scale clonal propagation of plants, but existing technical and biological problems prevent its use, at least in the immediate future (see Chapters 3, 4, 8, and 9, this volume).

The micropropagation industry started with orchids and ferns and is still largely restricted to horticultural plants, such as *Anthurium, Cattleya* and several other orchid species, *Ficus, Gerbera, Nephrolepis, Saintpaulia, Spathiphyllum,* and *Syngonium.* During the past decade, production of many bulb and tuber plants, such as lilies, potato, and cassava, has become quite common. Most importantly, many vegetables (cauliflower and its many relatives, leek, garlic and onion, lettuce, asparagus, cucumber), fruits (strawberry, kiwi fruit, raspberry, bananas, date palm), and a few spice and medicinal species (cardamom, pyrethrum) are now being propagated *in vitro* in sizable quantities. Finally, rapid progress has also been made in the micropropagation of woody species, which include many fruit (apple, plum, peach, cherry) and tree species (poplar, eucalyptus, willow, teak, sandalwood, and many conifers).

Ornamental plants still constitute more than 90% of all plants produced by micropropagation. The market for these plants is rather limited and is nearly saturated. There is thus little likelihood of any appreciable growth in the volume of plants produced by micropropagation in the future. As a result, the industry is rapidly becoming stagnant. In the past, the industry has been sustained by the comparatively high selling price ($0.30 or higher per plant) of *in vitro*–produced plants. Market forces and price structure, however, have forced the consolidation of the industry into a few large-volume producers. Many of the smaller operations, especially in the United States and Western Europe, where labor costs are particularly high, have not survived.

The high cost of production is directly related to the current labor-intensive methods of micropropagation. This has limited the range of species which are micropropagated *in vitro*. Estimates of labor costs run as high as 70–85% but never below 50%. The most expensive component of the system is the multiplication phase, which involves manual removal of the proliferating meristems from small individual culture vessels and the surgical separation and replanting of shoots in fresh media for further multiplication. Another labor-intensive part of the operation is the removal of plantlets formed *in vitro* and their replanting in soil for further growth in the greenhouse or the field. Automation of these components of the micropropagation technology can not only reduce the cost of plant production substantially but can also provide an economically attractive method for the scale-up of micropropagation in many vegetable and tree species where the market size is orders of magnitude larger than that for all of the ornamental species combined.

A variety of experimental strategies and systems to reduce labor costs by scale-up and automation of micropropagation are described and discussed in various chapters in this volume. Although most of these are not being used for commercial micropropagation yet, they do provide an insight into the contemporary thinking on this matter of critical importance. Several procedures, which focus on one or more phases of micropropagation, are currently under development and testing. These are:

- Computer programs for maintenance of database and record keeping of all micropropagation operations.
- Automated sterilization of primary explants.
- Automated preparation, sterilization, and dispensing of large volumes of nutrient media into individual containers.
- Improved culture vessels which allow greater gaseous exchange and improved light conditions.
- Growth of cultures on shallow layers of liquid media.
- Growth of cultures on inert supporting materials.
- Automated replenishing of nutrient media at preselected intervals.
- Growth of cultures under nutrient solution mist.
- Semiautomated systems to control nutrient supply and the gaseous environment for the growth and maintenance of shoot hedges which can serve as a source of shoots for rooting.
- Controlled gaseous environment and optimum light conditions to increase the efficiency of photosynthesis and photoautotrophic growth, resulting in the reduction or elimination of sucrose in the medium.
- Moving cultures and plants on conveyor belts.

- Use of robots to separate, cut, and reculture shoot cultures.
- Semi- and fully automated nodule culture in liquid media, with or without inert support matrices.
- The Vitromatic System, an integrated biological and mechanical system for micropropagation.
- *Ex vitro* rooting.
- Use of robots or automated systems for the transfer of *in vitro* plantlets to soil and the greenhouse.
- Mechanized field planting of tissue culture regenerants.
- Semi- and fully automated bioreactor systems for the production of somatic embryos and their encapsulation, with and without image analysis and sorting capabilities.

It is clear that further growth and expansion of the industry cannot be achieved without the inclusion of vegetable and forest species. The contemporary methods of micropropagation are totally inadequate for the scale-up of plant production to a level which will satisfy the large volume needs of the vegetable and forest industry. In order to enter this market, production costs must be reduced to 25% or less of the current levels, to less than $0.05 per plant. Such a drastic reduction in cost cannot be achieved without total automation of micropropagation, or at least automation of the most labor-intensive and expensive components of the system, namely the removal, separation, and replanting of the proliferating shoots and the removal of shoots/plantlets from culture and their planting into soil. Growing the shoots in individual containers under optimum light and temperature conditions during the multiplication and growth phases also adds substantially to production costs. Elimination or reduction of these steps can help further reduce production costs.

The recently developed Vitromatic System effectively addresses many of these problems, particularly those involving manual labor (Levin *et al.*, 1988; Levin and Vasil, 1989). It includes growing and multiplying large tight clusters of meristems in a liquid medium in automated bioreactors, which eliminates the need for repeated manual separation of meristems for multiplication, as well as the necessity to maintain large numbers of cultures in environmentally controlled growth rooms. Automation of this phase can also reduce the risk of infection, which is otherwise high during manual operations. At the appropriate time, the proliferating meristem clusters are automatically transferred to a bioprocessor, which cuts, sizes, separates, and distributes the resulting propagules into large capacity culture boxes with matrices, where they are allowed to grow into shoots or plantlets with roots. This process eliminates the second labor-intensive component of conventional micropropagation. Finally, an automated transplanting machine transfers the plants from matrices to greenhouse trays, thus eliminating the last laborious and time-consuming part of the process. Clearly, this integrated biological and mechanical system contains many of the elements necessary for increasing the efficiency and reducing the cost of micropropagation. The utility of the Vitromatic System has been demonstrated with a variety of plant species. With further refinements it has the potential to become a universally applicable method for the rapid and mass propagation of plants.

It is very likely that by the end of this decade most of the large commercial micropropagation operations will be partially or fully automated, utilizing one or more of the processes discussed above, resulting in the reduction of the cost of micropropagation to less than $0.05 per plant. Only then can this technology be used to generate the hundreds of millions of plantlets needed to support the transplant vegetable and forestry operations.

The production of vast numbers of clonal plants will be of great benefit to reforestation programs, which are urgently needed in many regions of Africa, Asia, Eastern Europe, and Latin America and can play a direct and valuable role in the improvement of the environment. High efficiency, low cost, and automated micropropagation can also make a significant contribution to the production of vegetable crops, where hybrid seed must otherwise be expensively produced each year, and for many other crops which are propagated vegetatively. Similar benefits can be derived in the production of spice and plantation crops, as well as fruit, spice, and medicinal plants.

REFERENCES

Levin, R., Gaba, V., Tal, B., Hirsch, S., DeNola, D., and Vasil, I. K. (1988). Automated plant tissue culture for mass propagation. *Bio/Technology* **6,** 1035–1040.

Levin, R., and Vasil, I. K. (1989). Progress in reducing the cost of micropropagation. *IAPTC Newsletter* **59,** 2–12.

Morel, G. (1960). Producing virus-free *Cymbydium. Amer. Orchid Soc. Bull.* **29,** 495–497.

Morel, G. (1965). Clonal propagation of orchids by meristem culture. *Cymbidium Soc. News* **20,** 3–11.

Morel, G., and Martin, C. 1952. Guerison de Dahlias atteintsd'une maladie a virus. *C.R. Acad. Sci., Ser. D* **235,** 1324–1325.

Sachs, T., and Thimann, K. V. (1964). Release of lateral buds from apical dominance. *Nature (London)* **201,** 939–940.
Skoog, F., and Miller, C. O. (1957). Chemical regulation of growth and organ formation in plant tissues cultured *in vitro. Symp. Soc. Exp. Biol.* **11,** 118–131.
Vasil, I. K., and Vasil, V. (1980). Clonal propagation. *In* "Perspectives in Plant Cell and Tissue Culture" (I. K. Vasil, ed.). *Int. Rev. Cytol. Supp.* **11A,** 145–173.

CHAPTER 2

Current Methods of Commercial Micropropagation

Sharon L. Kurtz
Robert D. Hartman
Irwin Y. E. Chu

Twyford International, Inc.
Santa Paula, California

I. INTRODUCTION

Worldwide, there have been numerous efforts to effectively commercialize the large-scale propagation of plants by tissue culture, and some of these efforts have been more successful than others (Chu and Kurtz, 1990). This chapter analyzes the currently available technology and how it is applied on a commercial level and identifies the areas in which important improvements can be made. The industry is at a critical stage now in that two major challenges must be effectively addressed before it can expand further: (1) reduction of current production costs, and (2) identification of key products which are differentiated by their high-value, high-volume nature. The unique pairing of these two considerations can synergistically expedite the industry's expansion and future scope. Before exploring current production costs, it is useful to review the advantages and limitations associated with this technology which have been identified by commercialization efforts to date. A thorough understanding of the current production systems is a prerequisite to studies aimed at cost reduction.

A. Advantages of Micropropagation

Micropropagation technology offers many unique advantages when compared with other more conventional propagation methods (George and Sherrington, 1984; Pennell, 1984; Murashige, 1974; 1978a,b). The majority of these advantages are associated with the development and marketing of improved products: (1) product development, (2) product enhancement, and (3) product marketability. These associated advantages can act to differentiate resulting products in the marketplace, as has been demonstrated most clearly for ornamental species and more recently for a food crop, such as the potato.

1. Product Development

Micropropagation can speed the development of improved products by speeding their release.

- Rapid multiplication. Micropropagation offers a method to rapidly increase both asexually propagated and sexually increased (via seed) materials. In the case of seed-produced hybrids, parent lines may be clonally increased to expedite the buildup of a large supply of hybrid seed. Of particular interest is the increase of male sterile lines, whose maintenance and increase normally require much effort and time in backcrossing. Many flower and vegetable seed companies have realized that by the rapid increase of seed parent lines, a hybrid variety can be released 3 to 5 years sooner than with seed-mediated parent stock (Kurtz, 1987).

- Product uniformity. The resulting product can have a high degree of phenotypic uniformity since the crop can be artificially manipulated in the laboratory to yield a large plant population at the same growth stage. In addition, genetic integrity can be maintained by utilizing axillary shoot or nodal multiplication systems. Resulting plant material can be delivered in a uniform product format, such as established plugs.
- High volume. Large populations can be produced in a relatively smaller growing space and in a reduced timeframe (Fig. 1). This benefit extends to the development of both sexually and asexually propagated varieties, since large volumes of the final selection or of the seed parent lines are required in each respective case.
- Elite selections. A phenotypically uniform population can be reproduced from a genetically heterozygous selection characterized by its uniquely desirable traits. Thus, it is possible to effectively capitalize on the selection of one desirable plant, even heterozygous in nature, which can be micropropagated to large numbers and released as a superior selection, bypassing conventional seed-mediated increases. This application could be particularly valuable for many tree and plantation crops which do not breed true to type from seed. Micropropa-

Fig. 1. Cultures are grown in an artificially controlled environment of temperature, light intensity and quality, photoperiod, and aseptic conditions. A range of 150–200 cultures can be grown per square foot of growth room shelf space.

gation permits the cloning of individual elite selections identified in a heterogeneous population to yield a population of genetically identical progeny.

- Genetically engineered products. Since many improved plant selections resulting from gene transfer programs are heterozygous or even sterile, micropropagation offers a unique means to expedite the commercialization of these materials via their rapid vegetative increase.
- Germ plasm storage. Storage methods for the effective preservation of valuable selections can be accomplished by combining micropropagation techniques with cold storage, minimal growth media, and even cryopreservation in liquid nitrogen. Seed companies have found that the *in vitro* maintenance of selected seed parent lines allows for their timely increase, since valuable time can be saved in the overall production cycle and large numbers of plants can be more readily generated on short notice. In addition, these methods, particularly cryopreservation, can be utilized to preserve the valuable genetic diversity of a crop's wild relatives, as well as species whose value has yet to be discovered. Many international germ plasm repositories have effectively integrated these techniques into their preservation efforts.

2. Product Enhancement

A product's quality, ultimate marketability, and thus commercial value can be effectively enhanced by the generation of products favorably characterized by their value-added features (Chu, 1986).

- Improved phenotype. The phenotype or appearance of the resulting product can be manipulated by adjusting the *in vitro* handling methods or environmental parameters, particularly growth regulators. Characteristics which can be manipulated include: growth habit, degree of basal branching, faster growth rate and/or flowering, and enhanced color and quality. For example, well-branched plants of *Ficus* and *Syngonium* can be produced by adjusting the growth regulators in the culture medium and varying the culture cutting method when the shoot clumps are divided.
- Disease indexed plants. Techniques to index or eliminate specific diseases, particularly viruses, can readily be incorporated into micropropagation procedures. This approach is particularly valuable when culture initiation is preceded by the necessary disease indexing and elimination by meristem culture. Commercial production efforts for both ornamental (e.g., lily, geranium) and food crop species (e.g., potato, strawberry) have benefited greatly from this application, which favorably distinguishes these products in the marketplace.

3. Marketability of Product

Marketing efforts can be enhanced via specific product formats and eased product flow on an international level.

- Product format. A diverse range of product formats can be used, such as rooted or unrooted microcuttings, bulblets, and liners established in either soil or artificial substrates (Fig. 2). In addition, *in vitro* cultures can be transported if a ready supply is required at another facility or company for rapid buildup.
- Product movement. Since plants can be produced certified free of specific pathogens and can also be shipped free of soil, international markets are more readily accessible. International phytosanitary regulations are more easily met by this means of plant production than by any other. This international product flow also supports the establishment of offshore production facilities to capitalize on lower labor costs and expanded markets.
- Nonseasonal production. Production efforts can proceed year-round since the multiplication phase is completed under artificially controlled conditions. Once culture requirements, multiplication rates, and acclimation and establishment procedures are defined for a given crop, production cycles can be scheduled to meet peak demands. In addition, the ability to cold store cultures contributes further to serving a greater number of markets, while easing production peaks.

B. Limitations of Micropropagation

While there are numerous advantages associated with micropropagation as a method of plant increase, three major limitations to its extended application exist: (1) product line restrictions, (2) customer acceptance, and (3) high production costs. Since the commercialization of this technology more than twenty years ago, numerous limitations were encountered and addressed. Problems with crop scheduling, inconsistent quality, seasonality, and inadequately tested protocols have historically been major problems, but tremendous progress has been made in overcoming these obstacles.

1. Product Line Restrictions

- Available micropropagation protocols. The choice of crops to be produced via micropropagation is limited to those species for which commercially feasible protocols have been defined. In spite of the fact that

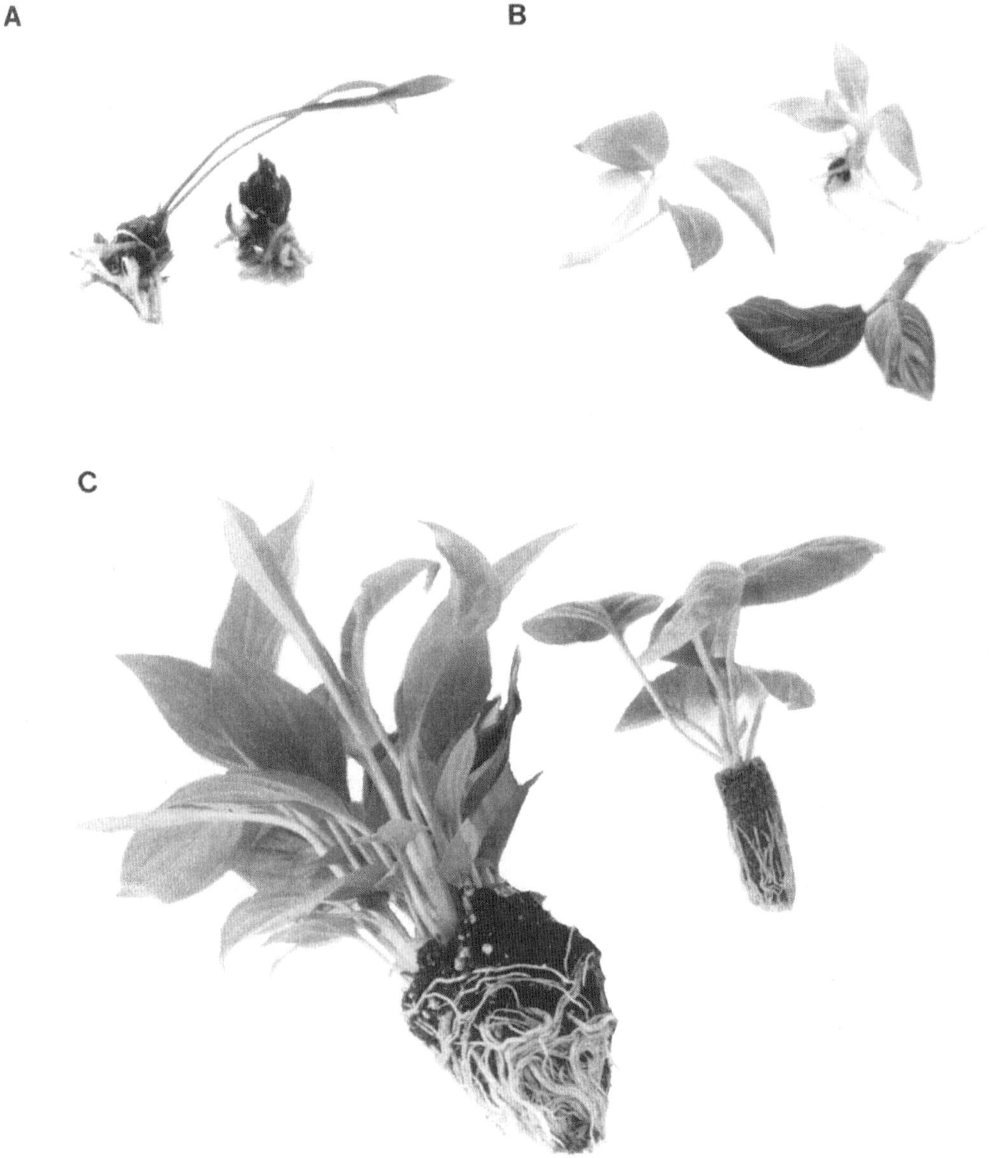

Fig. 2. Common product formats. (A) Lily bulblets. (B) Microcuttings, clockwise from left, *Dieffenbachia*, *Musa*, and *Calathea*. (C) Stage IV liners: *Spathiphyllum* soil-established liner (left), and *Gerbera* soilless plug (right).

there are numerous reports in the literature describing protocols for a wide range of plant species, they are often not suited to commercial scale-up (Giles and Morgan, 1987). A commercially feasible protocol must be highly reproducible and include acceptable procedures for culture initiation, rapid multiplication, maintenance of a high multiplication rate throughout numerous generations, successful rooting, acclimation, and establishment in soil.

- Market demand. The market demand will, of course, determine if the product formats and production costs associated with micropropagation are acceptable for specific crops. In addition, market demand will determine which potential products merit protocol development costs if acceptable systems are not currently available.

2. Customer Acceptance

- Product quality. Another major factor to be addressed is the consistent delivery of high quality microcuttings and plants. Ornamental species such as foliage plants and flowering pot plants are specifically evaluated and thus, priced, on the basis of every portion of the plant (Hartman and Zettler, 1986). Necessary quality control measures include: roguing off-types, grading by size, and product grow-outs. Genetic and phenotypic stability is an important concern. In most cases, however, such transient or permanent changes can be avoided or minimized by: (1) utilizing reduced growth regulator concentrations, (2) avoiding a callus intermediate and adventitious shoot regeneration, and (3) reducing subculture interval and duration in culture (Hussey, 1983).
- Product delivery. Customer acceptance is also associated with the ability to serve as a reliable producer who can schedule crops accurately and consistently deliver the product in a timely manner. Micropropagation is best suited to providing a steady stream of plant material instead of adapting to a customer's seasonal demand (Hartman, 1985). Under contract production, long-range planning is required of both purchaser and producer, since a purchase commitment of up to one year and a notice of at least three months before desired production schedule changes is required.

3. High Production Costs

Micropropagation production costs are high compared with those of conventional methods such as cuttings, division, or seeds (George and Sherrington, 1984) and are not cost competitive for many species. Crops which are currently micropropagated exhibit the numerous benefits stated earlier, which compensate for their generally higher unit costs. George and Sherrington (1984) outlined the following instances in which the cost of micropropagation could be viewed as acceptable: (1) a plant which cannot be propagated easily or quickly by other methods, (2) a high cost of conventional methods, (3) conventional methods which require specialized techniques (e.g., grafting), and (4) the capability of producing disease-free plants. An in-depth analysis of current commercial production is presented later in Section IV.

II. CURRENT PRODUCTION SYSTEMS

Commercial micropropagation laboratories utilize four basic methods to multiply plants *in vitro:* (1) enhanced axillary branching, (2) nodal sections, (3) adventitious shoots, and (4) somatic embryogenesis. Of these, somatic embryogenesis is used the least. The defined protocols available for a particular species, as well as other associated end-product benefits or risks, ultimately determine the best method for commercialization.

A. Enhanced Axillary Branching

The enhanced axillary branching method is probably the most frequently used of the four methods described. Both terminal and axillary shoot meristems are used to initiate cultures. By manipulation of the culture division technique and culture environment, particularly the growth regulators in the culture medium, apical dominance is suppressed and axillary branching is induced (Fig. 3A). Cultures may be divided as either single shoots or shoot clumps and further multiplied to obtain the desired number of shoots. Although the initial multiplication rate is slow, it increases rapidly and reaches a steady plateau for several subcultures. Excised shoots may then be rooted *in vitro* or directly in soil.

B. Nodal Sections

The nodal sections method is associated with the least amount of possible genetic variation, since a callus intermediate is avoided. Individual nodal sections with their associated axillary vegetative buds are excised upon subculture (Fig. 3B). The axillary bud grows and elongates, pro-

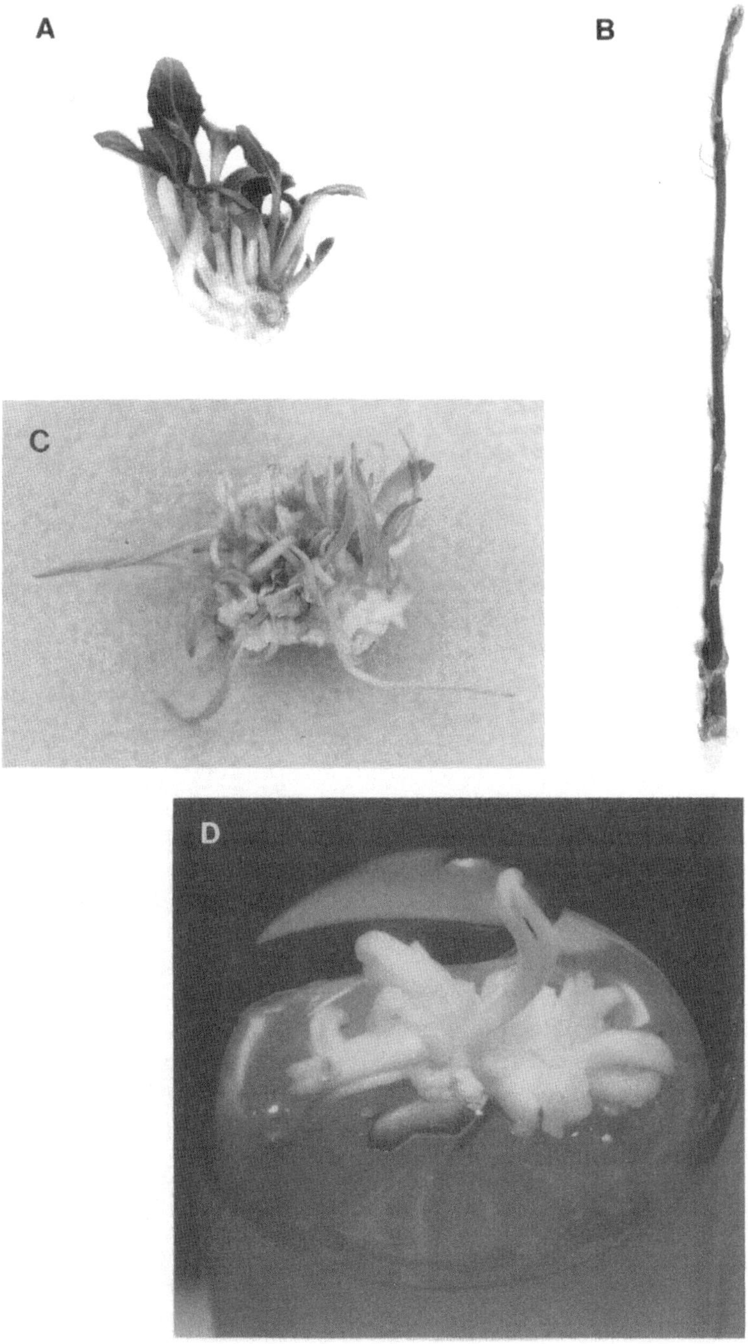

Fig. 3. Commercial micropropagation methods most commonly utilized. (A) Enhanced axillary branching, *Gerbera*. (B) Nodal sections, *Asparagus*. (C) Adventitious shoots, *Hemerocallis* (day lily). (D) Somatic embryogenesis, *Phoenix dactylifera* (date palm).

ducing additional nodes for excision. One- or two-node sections can be excised and rooted *in vitro* or directly in soil to produce complete plants. The only associated disadvantage with this method is that not all species can be propagated successfully by this method.

C. Adventitious Shoots

Adventitious shoots can be induced to form directly from such explants as stems or leaves, or even indirectly from a callus intermediate (Fig. 3C). Shoot-producing cultures can be subcultured until the desired number of shoots has been obtained. As in the other previously described methods, shoots may be excised and rooted *in vitro* or in soil. Of the four commercial micropropagation methods, this method is the least genetically conservative (Murashige, 1978b). Techniques to minimize variation were discussed earlier and should be taken into account when utilizing the adventitious shoot regeneration and multiplication method. This method, however, is still not recommended for the micropropagation of crops whose genetic integrity is of major concern, e.g. parent lines to be used in F_1 hybrid seed production.

D. Somatic Embryogenesis

Somatic or asexual embryos can be induced to differentiate directly or indirectly from the cells, organs, or intermediate callus of a plant. Most often, somatic embryos will further proliferate to yield secondary embryos (Fig. 3D). These embryos may be grown into complete plants and established in soil. Somatic embryogenesis is viewed as having the potential to be the most efficient plant regeneration method, since literally millions of plants could be produced from a relatively small amount of cultured cells or tissue. Significant research efforts continue to be directed toward the utilization of this regeneration pathway to produce artificial seeds via the coating or encapsulation of somatic embryos (Redenbaugh *et al.*, 1986, 1987; Lutz *et al.*, 1985). Additional research is still required in the areas of embryo encapsulation, storage, and conversion to plants. The major limitation of somatic embryogenesis as a regeneration method for commercial plant production is that very few species have been adequately characterized concerning the development of commercially feasible protocols.

III. PRODUCTION COST COMPONENTS

A. Importance of Understanding Production Costs

The importance of being able to accurately understand and determine micropropagation production costs cannot be overemphasized, due to its direct linkage to financial profitability. Although profitability is only one method of measuring performance, it is still the best way to determine the relative production/selling cost advantages associated with a given product or product line. The information required to formulate a workable production plan, such as culture or shoot multiplication rate, finishing time, percent loss, as well as space and labor requirements, directly impacts on cost determinations. Once the prerequisite crop data have been collected for crop scheduling, they can be utilized in combination with other additional data to develop meaningful cost analyses. It is only after these calculations can be made that profit margins and, therefore, selling costs can be accurately assigned. This information is ultimately combined with market and sales data to determine which existing products should be continued and which new products under development should be commercialized. Production cost analysis should also be used in conjunction with the research and development of new micropropagation systems or methods to adequately evaluate their cost-effectiveness throughout their development. Obviously, a thorough understanding of production costs serves as a meaningful competitive advantage in the micropropagation industry.

B. Production Cost Determination Methods

Production costs can be determined by many methods. Donnan (1986) suggested calculating the cost of operating a transfer hood by dividing the total operating costs by the number of transfer hood positions and then multiplying by the number of hours per position for the time in question, thus yielding the cost of operating one transfer hood position for 1 hour. It was further suggested that the particular crops being handled and their associated work rate should then be examined to determine whether a profit is being realized. Strain (1980, 1981) recommended determining production costs by calculating the costs per square foot of shelf space. Donnan (1986) reported an average cost of $1.75 per square foot per week. This calculation emphasizes the importance of maintaining growth room shelves at full capacity. To date, labor costs in the range of 40% (Sluis and Walker, 1985) and 60% (Donnan, 1986) to more than 70% (Anderson and Meagher, 1977) have been reported.

In our own experience, the most accurate method of production cost analysis focuses on a crop-by-crop determination, whereby detailed records are maintained on a daily basis concerning: (1) culture division and transfer rates, (2) number of cultures per vessel, (3) number of vessels, (4) culture multiplication rates, (5) losses due to contamination, poor cutting technique or poor clone performance, (6) space allocations, and (7) total plant or microcutting population desired.

C. Determining Laboratory Cost Components

If only microcuttings are produced, then the entire production cost analysis effort must focus solely on laboratory-based production components, whereas if soil-established liners are to be produced, then both laboratory and greenhouse-based cost components must be characterized. To thoroughly understand the various laboratory cost components of a micropropagation system, it is useful to review the four basic production stages recognized by Murashige (1974, 1977, 1978a,b). This classification of production stages has been extensively adopted and utilized in commercial micropropagation (Fig. 4). Stages I–III occur *in vitro,* while Stage IV generally occurs in a greenhouse environment. Deberg and Maene (1981) suggested an additional stage referred to as Stage 0. The adoption of such a system greatly simplifies not only such internal functions as production planning, accounting, and cost analysis, but also eases external communication with customers and other laboratories.

1. Stage 0

Stock plants used for initiating cultures are grown in a relatively clean environment to reduce threats of both disease and contamination. Disease indexing procedures can be implemented to certify the stock plant's freedom from specific pathogenic viruses, bacteria, and fungi (Lawson, 1986). This effort has been successfully implemented in the commercial

Fig. 4. Production stages for the commercial micropropagation of *Ficus elastica* by enhanced axillary shoot multiplication. (A) Stage I, excised shoot tip (0.5 cm). (B) Stage II, multiplying shoot culture (2.0 cm). (C) Stage III, rooted shoot clump (3.0 cm). (D) Stage IV, finished plant grown as a liner (7.0 cm from the soil).

A
B
C
D

micropropagation of many ornamental species, e.g., foliage plants, carnations, chrysanthemums, and geraniums (Knauss, 1976), as well as for such agriculturally important crops as potato and strawberry (George and Sherrington, 1984). Viruses may be detected by: (1) indicator plants, (2) enzyme-linked immunosorbent assay (ELISA), (3) electron microscopy, (4) cDNA probes, or (5) double-stranded RNA (dsRNA) for RNA viruses. Viroids may be detected by polyacrylamide gel electrophoresis (PAGE) and cDNA probes. Methods for detecting bacteria include: (1) selective culture media, (2) physical and physiological characteristics, and (3) serological methods such as tube agglutination and precipitation, immunofluorescence, and ELISA (Lawson, 1986). Fungal pathogens are generally identified by: (1) growth on selective culture media, (2) physical and physiological characteristics, and (3) serological techniques such as ELISA and immunofluorescence.

Of course, the specific crop to be produced will partially determine which, if any, disease indexing methods are required. In many cases, the utilization of disease indexing and certification techniques has resulted in expediting international exchange of germ plasm and products. As competition increases not only between various commercial laboratories, but also with more conventional propagation methods, the "value-added" component of disease elimination in the production process assumes greater importance. Most generally, this cost component is isolated, or managed apart from the other stages. Disease indexing is not always included in the commercial production system unless a specific customer requests it, or if certified disease-free plants or microcuttings are to be specifically produced, as would be the case for a crop such as potato.

2. Stage I

During Stage I, an aseptic culture is initiated and established (Fig. 4A). Terminal as well as axillary shoot meristems generally serve as the explant source. Possible bacterial and fungal contamination must be detected during this stage to prevent large crop losses later in the production cycle. If virus elimination is required, meristem culture is performed at this time. The small meristematic dome of a shoot, 0.1–0.5 mm high, is excised and cultured. In some cases, heat treatment and/or chemotherapy may also be required before meristem culture. Although smaller meristems are generally associated with greater potential virus elimination, such cultures are also susceptible to possible tissue culture–induced variation since a callus intermediate may result and/or adventitious shoots may regenerate. The virus indexing methods discussed earlier would be utilized to evaluate the virus-fee status of the cultures and resulting plants.

Each species has its own associated requirements and production processes, making the inclusion of Stage I costs in a production cost analysis rather nonroutine. It is often best to isolate this initiation cost apart from the more meaningful Stage II–IV costs. On a per-unit basis, Stage I definitely represents the most costly portion of the entire process. Individual shoots or meristems are cultured in single test tubes, and cultures must be carefully monitored for their possible contamination and growth response. The most meaningful way to reduce Stage I costs is to grow stock plants (Stage 0) in a clean environment. Meristem culture requirements can be reduced if certified virus-free stock can be located and purchased for culture initiation. In general, Stage I requires from 3 to 24 months, with at least four subcultures.

3. Stage II

During Stage II, shoots, in the case of axillary branching, nodal, and adventitious shoot systems, are multiplied on a regular basis (Fig. 4B). In the case of somatic embryogenesis, embryos are routinely produced and rapidly increased. A subculture and/or harvest cycle of 4 weeks with a twofold to sevenfold increase is generally the rule for most shoot multiplication systems. Once enough shoots have been obtained, individual shoots may be harvested as unrooted Stage II microcuttings, or they may be carried on to Stage III and rooted. Stage II can last 10–36 months, depending on the species and its associated ability to maintain acceptable multiplication rates and exhibit minimal variation and off-types.

Stage II represents one of the most costly stages of the production cycle, since repeated labor is required on a regular basis. Commercial laboratories try to maximize the total plant or microcutting output to minimize these costs over time. The production cost on a per unit basis declines and reaches a relatively stable level once multiplication rates increase and stabilize (Henny *et al.*, 1981). The benefit of extending the length of time a specific culture initiate is maintained in culture must be weighed against the associated possible risks of contamination or off-types. This may vary from crop to crop. During Stage II, it is necessary to keep records of each clone's performance at every subculture, including: (1) multiplication rate, (2) labor requirements such as culture division and transfer rates, and (3) percent loss due to poor technique, contamination, or clonal response. These records, which should be kept for each transfer technician, also serve as an excellent method of employee performance appraisal. At Twyford, a detailed computerized inventory control system utilizes these data to accurately forecast product availability on a weekly basis.

4. Stage III

For most crops, individual shoots or shoot clumps are next rooted *in vitro*, necessitating another subculture and additional handling in the laboratory (Fig. 4C). For some crops, Stage III may also include a shoot development or elongation phase, whereby the shoots in Stage II are multiplied as a lawn of very short shoots which must be elongated before rooting. Depending on the species, shoots produced during Stage II may be transferred directly to soil for rooting in a greenhouse. Other alternatives include rooting the excised shoots in artificial substrates such as Oasis™ foam or rockwool. Depending on the species, Stage III generally requires 1–6 weeks.

While Stage III requires only one subculture, it does require additional labor and an additional type of culture medium, thus further increasing production costs. As indicated earlier, for some species it is possible to bypass Stage III and proceed directly to Stage IV, thus impacting positively on production cost.

5. Relative Laboratory Cost Components

Stages 0–III, with the primary focus on Stages II and III, can be evaluated to determine the relative contribution of the laboratory-associated portions of the commercial micropropagation process. Table I indicates that labor constitutes 64% of the costs, with 34% represented by the transfer technicians. As stated earlier, transfer technicians spend most of their time dividing and transferring Stage II cultures (Fig. 5). Thus, this calculation clearly indicates the need for labor reduction efforts on the part of Stage II culture handling methods.

D. Determining Greenhouse Cost Components

Greenhouse-associated costs must be included in cost analyses involving soil-established liners or plugs. Of the total production costs for the

TABLE I
Relative Laboratory Production Cost Components[a]

Cost components	Percentage
Direct labor	34
Indirect labor (media preparation, initiation, etc.)	15
Supervision	15
Total labor	64
Utilities	10
Depreciation	12
Other production costs	14
Total	100

[a]Twyford International, Inc., Santa Paula, California, operation study, 1989. These calculations are based on a total annual production volume of 7.5 million units. A total of 12 different plant genera and 70 different clones were produced.

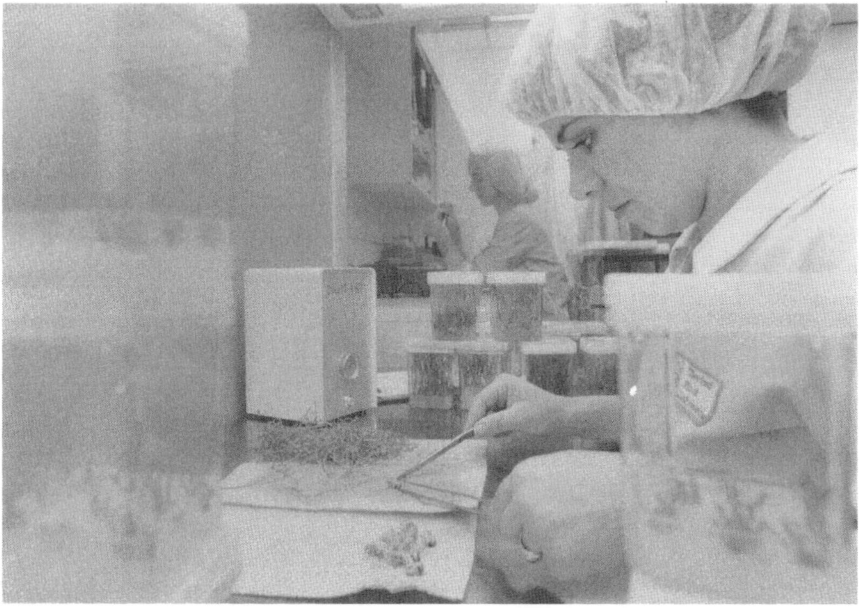

Fig. 5. Currently, all Stage II cultures must be divided manually using scalpels and forceps. Whether during commercial production or during the development of micropropagation systems, technicians must record such information as cutting rate per hour and note any problems with clone performance.

TABLE II
Relative Greenhouse Production Cost Components[a]

Cost components	Percentage
Planting direct labor	11
Indirect labor (irrigation, maintenance, sorting, dispatch)	30
Supervision	9
Total labor	50
Utilities	8
Depreciation	12
Other production costs	30
Total	100

[a]Twyford International, Inc., Santa Paula, California, operation study, 1989. These calculations are based on a total annual production volume of 7.5 million units. A total of 12 different plant genera and 70 different clones were produced.

Twyford International study described in Tables I and II, laboratory costs represented 60%, while greenhouse costs were 40%.

1. Stage IV

Special care is necessary to ensure the successful transfer of Stage II or Stage III shoots and/or shoot clumps to the *in vivo* greenhouse environment to produce soil-established liners (Fig. 4D). Several techniques have been developed to ease the acclimation process, including: (1) an artificial acclimation room with controlled environmental conditions, (2) plastic-covered incubators in the greenhouse, (3) plastic tents in the greenhouse, and (4) a dense, fine particle fog system in the greenhouse. Of these methods, plastic tents and fog are the two most widely employed acclimation methods in the industry (Fig. 6). Currently, cultures are sorted and planted manually in the greenhouse.

2. Relative Greenhouse Cost Components

When analyzing the costs associated with Stage IV, it is important to collect and maintain detailed records concerning: (1) planting rate, (2) percent survival, (3) sorting rate, and (4) packing rate. The relative greenhouse cost components can then be determined. Table II indicates that labor constitutes 50% of the total greenhouse costs, with 11% represented by the initial planting process and 30% contributed by labor re-

Fig. 6. The production of Stage IV liners requires careful attention to greenhouse acclimation. Large plastic tents, as shown above, or dense fog are the most commonly used acclimation methods.

quired for irrigation, maintenance, sorting, and dispatch. Obviously, the relative total labor requirement is much higher in the laboratory (64%) than in the greenhouse (50%).

IV. DEVELOPMENT OF COST-EFFECTIVE PRODUCTION SYSTEMS

A. Importance of Cost-Reduction Strategies

As mentioned in the beginning of this chapter, the future expansion of the micropropagation industry to include a greater variety of plant species, such as plantation, forest, and vegetable crops, depends upon the reduction of current production costs and ultimate selling prices. More cost-effective production systems will support competitive pricing compared with the more conventional propagation methods of cuttings or seed. Of course, the greater the associated benefits of producing a particular species via micropropagation, the greater the relative selling cost accepted by the marketplace. Commercial micropropagation laboratories have long been engaged in price wars to sell the lowest-priced product, even when it meant that the company did not profit or even did not cover expenses. Since there is very little product diversity in the micropropagation industry, competition for a limited number of specialized markets continues to intensify. Consequently, there has been a rapid decrease in the number of smaller laboratories over the past few years, with larger laboratories capturing the limited market (Rowe, 1986; Harrell and Simonton, 1986).

B. General Cost-Reduction Strategies

Three general strategies to reduce current production costs include: (1) *in vitro* biological optimization, (2) elimination of production stages or repeated subcultures, and (3) automation and mechanization. Throughout the development of a commercially feasible micropropagation production system, it is very important to address each of these strategies simultaneously whenever possible. The ultimate result of this approach will be the realization of the most cost-effective system possible in the shortest amount of development time.

1. *In Vitro* Biological Optimization

Although substantial effort has been directed toward the optimization of *in vitro* responses for plant cultures, significant opportunities remain for further increases in not only yield, i.e., the number of plants or microcuttings, but also in overall product quality. In terms of ultimate product quality, major concerns include the minimization of off-types, vitrification, and contamination. The acclimation and establishment of Stage II or III plant material with a very low degree of loss must be an ultimate goal. Efforts to optimize the biological components of production systems include addressing both chemical and physical parameters associated with the *in vitro* culture environment.

- Chemical parameters. The major factors worth in-depth consideration include: the inorganic and organic components of the culture media, the types and concentrations of growth regulators utilized, culture

medium pH, the possible production of toxic by-products by the cultures themselves, and the gaseous phase, e.g., ethylene generation or CO_2 supplementation.
- Physical parameters. The two major factors which merit further study include: temperature and the quality, intensity, and duration of light.

2. Elimination of Production Stages or Repeated Subcultures

Any effort to simplify the production system, most generally through the reduced number of different production stages, or by maximizing the yield of a given production stage should be evaluated in more depth concerning their relative cost savings. Commercial laboratories are directing particular effort toward: the elimination of Stage III, reduced Stage IV requirements, and the large-scale expansion of Stage II with significantly reduced subculture requirements.

- Elimination of Stage III. As previously discussed, Stage III represents the *in vitro* rooting phase. For many species, this laboratory-based stage can be bypassed via the direct rooting of shoot clumps or microcuttings in soil or artificial substrates. Since a laboratory environment with its associated high costs for maintaining an aseptic, climatically controlled environment is no longer required for 1–6 weeks, depending on the crop, significant cost savings can be realized. The resulting available laboratory space can then be devoted to Stage I and II cultures. While Stage III has been successfully eliminated for many species, much additional research is required to extend this approach to a greater range of species.
- Reduced Stage IV requirements. Stage IV requirements, particularly those associated with acclimation, can be reduced through the careful manipulation and handling of Stage II or III cultures. It is possible to produce hardier shoots or plantlets which do not require such highly specialized acclimation environments, thus reducing handling and equipment costs.
- Stage II expansion and reduced subculture requirements. Since the majority of the commercial production effort is spent on the multiplication of Stage II cultures, any efforts to maximize Stage II yield and reduce the number of subcultures required would be very worthwhile. To date, major focus has been placed on the utilization of larger culture vessels with greater associated culture densities, as well as the use of liquid media to ease subculture efforts and possibly increase ultimate

product yield through associated biological optimization. These efforts will be explored in more depth in the next section on automation and mechanization.

3. Automation and Mechanization

Recently, much effort has been directed toward the application of automation and mechanization technology in the commercial micropropagation industry, as this volume seeks to present. The current objective of increasing total production and reducing associated costs to capture ever greater market shares cannot be met with current production technology (Sluis and Walker, 1985; Harrell and Simonton, 1986; Levin and Vasil, 1989; Chu and Kurtz, 1990). Automation and mechanization could be applied in almost all phases of commercial micropropagation in the form of computerization, materials handling, and robotics. Significant cost savings can be realized in both the laboratory- and greenhouse-based production phases.

- Laboratory phases. The major laboratory tasks for which various degrees of mechanization already exist include: (1) culture media preparation, (2) growth room management and materials handling, and (3) dishwashing. Much of this technology has already been developed for food-related industries, and it has only to be applied to micropropagation. Often, it is even possible to purchase equipment that is already available or which requires only minor modification.

  More complex laboratory tasks which still require automation include: (1) culture division and transfer, and (2) sorting and grading of Stage II microcuttings. As stated earlier, human labor costs currently represent up to 64% of total laboratory-related production costs. Of this total, the technicians responsible for culture division and transfer represent the majority of total labor requirements, i.e., 34%. Since most of this labor is dedicated to Stage II culture handling, most attempts to employ robotics in micropropagation processes have focused on the automated division and transfer of axillary or adventitious shoot clumps (Deleplanque *et al.,* 1985; Miwa, 1987a,b,c; Anonymous, 1988). In the case of nodal systems, few efforts have been made (Rowe *et al.*, 1987). In most cases, computer vision systems and a CCD camera have been incorporated to locate cultures and identify desired cuts. In addition to robotic systems, mechanized shoot multiplication systems utilizing liquid culture media in combination with solid culture media (Debergh and Maene, 1981; Maene and Debergh,

1985; Aitken-Christie and Jones, 1987), or even in place of solid media (Tisserat and Vandercook, 1985, 1986; Weathers and Giles, 1988) have been developed. To our knowledge, however, none of these systems has been commercially practiced. Most notable of liquid media-based culture systems developed to date is the Vitromatic System developed by Levin *et al.* (1988), whereby either shoot organogenesis or somatic embryogenesis may be employed for regeneration, depending on the species. Much effort has also been directed toward the large-scale production of artificial seeds through the encapsulation of somatic embryos (Styer, 1985; Redenbaugh *et al.*, 1987). Currently, computer vision systems are being developed to automatically sort, size, and grade somatic embryos according to size (Grand d'Esnon *et al.*, 1987). Other efforts to automate or mechanize portions of commercial micropropagation systems have been directed toward large-scale sterilization of explants (Maurice *et al.*, 1985) and acclimation chambers (Kozai *et al.*, 1987). It could also be possible to automatically harvest individual shoots or microcuttings.

The greatest obstacle to be overcome in the case of axillary or adventitious shoot multiplication systems is that of the complex growth habit of the cultures which could be difficult to separate by mechanical dissection (Rowe, 1986). It may not be practical to attempt the mechanical dissection of particular species which require a high degree of precision and expertise in their handling, such as is the case with *Gerbera*. In addition, current robot working speeds are still too slow compared with experienced technicians. Ultimately, the high investment cost to develop an automated system for division, transfer, harvest, and even planting, would need to be offset on the basis of the chosen target crops, realized increase in plant numbers, potential new or expanded markets, and ultimate return on investment (Chu and Kurtz, 1990).

- Greenhouse phases. The major greenhouse tasks for which directly applicable mechanization technology already exists include: (1) filling and dibbling growing flats, (2) transport of planted flats, and (3) irrigation and fertilization. This technology has been developed for the horticultural bedding plant industry. Equipment applicable to bedding plant production can be directly utilized in the finishing of tissue cultured liners. Of particular use will be the technology being developed and perfected for the plug-growing industry (Koranski and Laffe, 1985) and the automated greenhouse technology under development and commercialization, including palletized benching, computerized environmental control systems, and robotic handling systems.

Specific tasks which require further development include: (1) sorting and grading microcuttings or plantlets, (2) planting microcuttings or plantlets, and (3) grading finished plants. As discussed earlier, human labor costs make up 50% of total labor requirements, with 11% devoted to planting, 30% devoted to irrigation, maintenance, sorting, and dispatch, and 9% due to supervision. Currently, much progress is being made in the development of both automated transplanting technology and plant grading and sorting technology as applied to the plug production industry. In regard to automatic transplanting of seedlings, both machines (Kuack, 1989; Brewer, 1986; Levin *et al.*, 1988; Levin and Vasil, 1989) and robots (Kutz, 1986, 1987; Ting and Giacomelli, 1988) are being compared. In addition, other efforts are specifically focusing on the grading, trimming, and sticking of cuttings (Simonton and Pease, 1989). In addition, a few horticultural engineering companies are developing sophisticated machines to individually grade seedling plugs using complex computer vision systems which can not only measure height but can also compute total leaf area.

The greatest challenge to be faced in the application of current grading, sorting, and planting technology in micropropagation systems is that of the small scale and delicacy of the microcuttings involved, particularly in the handling and planting phases. Of course, the current costs associated with such machinery are still relatively high, e.g., $40,000 to $150,000 for an automatic transplanter. As technology is perfected, however, costs should decline.

V. CONCLUSIONS

Once the two major challenges of reducing current production costs and identifying additional products differentiated by their high-value, high-volume nature are successfully addressed, the plant micropropagation industry will expand to include an ever wider range of species. Major methods to reduce current production costs include both *in vitro* growth optimization and mechanization and automation. Numerous reports, including the material presented in this volume, support the premise that a major portion of micropropagation systems will become automated in the not-too-distant future. Throughout the development of any cost-reduction strategy, however, the return on initial investment must be considered. In addition, cost reduction must be understood and carefully analyzed throughout the application of either growth optimization or automation to yield the most cost-effective systems, while maintaining high product quality.

ACKNOWLEDGMENTS

The authors wish to thank Peter Tolley and Bill Boyd for excellent assistance with the photographs. We also wish to express our appreciation to our other colleagues at Twyford, including Peter Christl, Greg Lloyd, and Victoria Escoto for their valuable assistance in preparing this manuscript.

REFERENCES

Aitken-Christie, J., and Jones, C. (1987). Towards automation: Radiata pine shoot hedges *in vitro*. *Plant Cell Tissue Organ Cult.* **8,** 185–196.

Anderson, W. C., and Meagher, G. W. (1977). Cost of propagating broccoli plants through tissue culture. *Hort Science* **12(6),** 543–544.

Anonymous. (1988). Toshiba develops prototype robot for future plant factories for tissue culture of young plants. *Toshiba News* May 6, pp. 1–3. Tokyo, Japan.

Brewer, H. L. (1986). Automated seedling production. *Acta Hort.* **187,** 53–59.

Chu, I. Y. E. (1986). The application of tissue culture to plant improvement and propagation in the ornamental horticultural industry. *In* "Tissue Culture as a Plant Production System for Horticultural Crops" (R. H. Zimmerman, R. J. Griesbach, F. A. Hammerschlag, and R. H. Lawson, eds.), pp. 15–33. Martinus Nijhoff, Boston.

Chu, I. Y. E., and Kurtz, S. L. (1990). Commercialization of plant micropropagation. *In* "Handbook of Plant Cell Culture" Vol. 5 (P. V. Ammirato, D. A. Evans, W. R. Sharp, and Y. P. S. Bajaj, eds.), pp. 126–164. McGraw-Hill Publishing Co., New York.

Debergh, P. C., and Maene, L.J. (1981). A scheme for commercial propagation of ornamental plants by tissue culture. *Sci. Hortic.* **14,** 335–345.

Deleplanque, H., Bonnet, P., and Postaire, J. G. (1985). An intelligent robotic system for *in vitro* plantlet production. *In* "Proc. 5th Intl. Conf. Robot Vision and Sensory Controls" (N. J. Zimmerman, ed.), pp. 305–314. Amsterdam, Netherlands.

Donnan, A., Jr. (1986). Determining and minimizing production costs. *In* "Tissue Culture as a Plant Production System for Horticultural Crops" (R. H. Zimmerman, R. J. Griesbach, F. A. Hammerschlag, and R. H. Lawson, eds.), pp. 167–173. Martinus Nijhoff Publishers, Boston.

George, E. F., and Sherrington, P. D. (1984). "Plant Propagation by Tissue Culture: Handbook and Directory of Commercial Laboratories." Eastern Press, Reading, United Kingdom.

Giles, K. L., and Morgan, W. M. (1987). Industrial-scale plant micropropagation. *Tibtech* **5,** 35–39.

Grand d'Esnon, Harrell, R., and Chee, R. (1987). Propagule sorting by computerized vision during artificial seed production. *Proc. Moet-Hennessy Conf. Electronics Management Living Plants,* p. 22, Monte Carlo.

Harrell, R., and Simonton, W. (1986). Automation opportunities in plant tissue culture operations. *Amer. Soc. Agric. Eng.* Paper No. 86–1596.

Hartman, R. D. (1985). How tissue culture makes better crops. *In* "Ball Redbook" 14th ed. (V. Ball, ed.), pp. 165–172. Reston Publishing, Reston, Virginia.

Hartman, R. D., and Zettler, F. W. (1986). Tissue culture as a plant production system for foliage plants. *In* "Tissue Culture as a Plant Production System for Horticultural Crops" (R. H. Zimmerman, R. J. Griesbach, F. A. Hammerschlag, and R. H. Lawson, eds.), pp. 293–299. Martinus Nijhoff, Boston.

Henny, R. J., Knauss, J. F., and Donnan, A., Jr. (1981). Foliage plant tissue culture. *In* "Foliage Plant Production" (J. N. Joiner, ed.), pp. 137–178. Prentice-Hall, Englewood Cliffs, New Jersey.

Hussey, G. (1983). *In vitro* propagation of horticultural and agricultural crops. *In* "Plant Biotechnology" (S. H. Mantell and H. Smith, eds.), pp. 111–138. Press Syndicate of the University of Cambridge, Cambridge, United Kingdom.

Knauss, J. F. (1976). A tissue culture method for producing *Dieffenbachia picta. Proc. Fla. State Hort. Soc.* **89,** 293–295.

Koranski, D., and Laffe, S. R. (1985). Plug production. *In* "Bedding Plants III: A Penn. State Manual," p. 126–140. Penn. State Univ. Press.

Kozai, T., Hayashi, M., Hirosawa, Y., Kodama, T., and Watanabe, I. (1987). Environmental control for acclimation of *in vitro* cultured plantlets. (1) Development of the acclimatization unit for accelerating the plantlet growth and the test cultivation. *J. Agr. Met.* **42(4),** 349–358.

Kuack, D. L. (1989). Mechanical transplanter makes its debut. *Greenhouse Grower* **7(2),** 76–78.

Kurtz, S. L. (1987). Plant improvement: The combined roles of biotechnology and breeding. *Greenhouse Grower* **5(3),** 62–63, 66–69.

Kutz, L. J., Miles, G. E., and Hammer, P. A. (1986). Robotic transplanting of bedding plants. *In* "Agri-Mation 2" *Proc. Agri-Mation 2 Conf.,* (March 3–5, 1986), Chicago, Ill., pp. 78–90. Am. Soc. Agric. Eng. and Soc. Manuf. Eng.

Kutz, L. J., Miles, G. E., Hammer, P. A., and Krutz, G. W. (1987). Robotic transplanting of bedding plants. *Trans. of the ASAE* **30(3),** 586–590.

Lawson, R. H. (1986). Pathogen detection and elimination. *In* "Tissue Culture as a Plant Production System for Horticultural Crops" (R. H. Zimmerman, R. J. Griesbach, F. A. Hammerschlag, and R. H. Lawson, eds.), pp. 97–117. Martinus Nijhoff Publishers, Boston.

Levin, R., Gaba, V., Tal, B., Hirsch, S., DeNola, D., and Vasil, I. K. (1988). Automated plant tissue culture for mass propagation. *Biotech* **6(9),** 1035–1040.

Levin, R., and Vasil, I. K. (1989). Progress in reducing the cost of micropropagation. *IAPTC Newsletter* **59,** 2–12.

Lutz, J., Wong, J., Rowe, J., Tricoli, D., and Lawrence, R., Jr. (1985). Somatic embryogenesis for mass cloning of crop plants. *In* "Tissue Culture in Forestry and Agriculture" (R. Henke, K. Hughes, M. Constantin, and A. Hollaender, eds.), pp. 105–116. Plenum Press, New York.

Maene, L., and Debergh, P. (1985). Liquid medium additions to established tissue cultures to improve elongation and rooting *in vivo*. *Plant Cell, Tissue Organ Cult.* **5(1)**, 23–33.

Maurice, V., Vandercook, C. E., and Tisserat, B. (1985). Automated plant surface sterilization system. *Physiol. Veg.* **23(1)**, 127–133.

Miwa, Y. (1987a). Plant tissue culture and its automation. *J. Jap. Soc. Prec. Eng.* **53(2)**, 208–209.

Miwa, Y. (1987b). Plant tissue cultivator robot. *J. Soc. Inst. Control Eng.* **26(3)**, 219–220.

Miwa, Y. (1987c). Plant tissue culture robot operated by a shape memory alloy actuator and a new type of sensing. *Proc. Moet-Hennessy Conf. Electronics and Management of Living Plants*, p. 52, Monte Carlo.

Murashige, T. (1974). Plant propagation through tissue cultures. *Ann. Rev. Plant Physiol.* **25**, 135–166.

Murashige, T. (1977). Clonal crops through tissue culture. *In* "Plant Tissue Culture and Its Bio-technological Application" (W. Barz, E. Reinhard, and M. H. Zenk, eds.), pp. 392–403. Springer-Verlag, New York.

Murashige, T. (1978a). Principles of rapid propagation. *In* "Propagation of Higher Plants Through Tissue Culture: A Bridge Between Research and Application" (K. W. Hughes, R. Henke, and M. Constantin, eds.), pp. 14–24. U.S. Dept. of Energy, Washington, D.C.

Murashige, T. (1978b). The impact of plant tissue culture on agriculture. *In* "Frontiers of Plant Tissue Culture 1978" (T. A. Thorpe, ed.), pp. 15–26. Univ. of Calgary Press, Calgary, Canada.

Pennell, D. (1984). The impact of micropropagation on commercial production. *Sci. Hortic.* **35**, 28–33.

Redenbaugh, K., Paasch, B., Nichol, J., Kossler, M., Viss, P., and Walker, K. (1986). Somatic seeds: Encapsulation of asexual plant embryos. *Biotech.* **4(9)**, 797–801.

Redenbaugh, K., Viss, P., Slade, D., and Fujii, J. A. (1987). Scale-up: Artificial seeds. *In* "Plant Tissue and Cell Culture" (C. E. Green, D. A. Somers, W. P. Hackett, and D.D. Biesboer, eds.), pp. 473–493. Alan R. Liss, New York.

Rowe, W. J. (1986). New technologies in plant tissue culture. *In* "Tissue Culture as a Plant Production System for Horticultural Crops" (R. H. Zimmerman, R. J. Griesbach, F. A. Hammerschlag, and R. H. Lawson, eds.), pp. 35–51. Martinus Nijhoff.

Rowe, W. J., Rowe, N. M., and Roepe, H. P. (1987). The application of robotics for an automated tissue transfer system. *In* "Proc. Moet-Hennessy Conf. Electronics Management Living Plants," p. 41, Monte Carlo.

Sluis, C. J., and Walker, K. A. (1985). Commercialization of plant tissue culture propagation. *Newsletter Intl. Assoc. Plant Tissue Cult.* **47**, 2–12.

Strain, J. R. (1980). Analyzing costs in tissue culture laboratories. Staff paper 167, Food and Resource Economics Dept., Inst. Food and Agr. Sci., Univ. of Florida, Gainesville, Florida.

Strain, J. R. (1981). A cost accounting model for plant tissue culture laboratories. Unpublished summary of a presentation to the Tissue Culture Working Group of the Am. Soc. Hortic. Sci. meeting (Aug. 11), Atlanta, Georgia.

Styer, D. J. (1985). Bioreactor technology for plant propagation. *In* "Tissue Culture in Forestry and Agriculture" (R. R. Henke, K. W. Hughes, M. J. Constantin, and A. Hollaender, eds.), pp. 117–130. Plenum Press, New York.

Ting, K. C., and Giacomelli, G. A. (1988). A robot for transplanting plugs. *Greenhouse Grower* **6(12)**, 58, 60.

Tisserat, B., and Vandercook, C. E. (1985). Development of an automated plant culture system. *Plant Cell, Tissue Organ Cult.* **5(2),** 107–117.

Tisserat, B., and Vandercook, C. E. (1986). Computerized long-term tissue culture for orchids. *Am. Orchid Soc. Bull.* **55(1),** 35–42.

Weathers, P. J., and Giles, K. L. (1988). Regeneration of plants using nutrient mist culture. *In Vitro* **24(7),** 727–732.

CHAPTER 3

Synthetic Seed Technology

Keith Redenbaugh
Jo Ann Fujii
David Slade

Calgene, Inc.
Davis, California

I. INTRODUCTION

A. Invention of Synthetic Seeds

The origin of the concept of synthetic seeds was reviewed earlier (Redenbaugh *et al.*, 1987b), with Murashige attributed as the first to publicly discuss the theoretical concept in 1977 but without any experimental data (Murashige, 1977). The first constructions of synthetic seeds involved multiple-embryo coating of desiccated celery somatic embryos (Kitto *et al.*, 1982) and single-embryo encapsulation of hydrated alfalfa somatic embryos (Redenbaugh *et al.*, 1984). The necessary components of synthetic seeds depend on whether they are hydrated or desiccated (Table I) and have been discussed in several papers (Durzan, 1985; Fujii *et al.*, 1987a; Redenbaugh *et al.*, 1986, 1987b, 1988b; Styer, 1985).

B. Definition of Synthetic Seeds

Several reviews on somatic embryogeny have emphasized using somatic embryos in the form of synthetic seeds for plant propagation. Four types of synthetic seeds have been proposed (Fujii *et al.*, 1986, 1987a,b):

TABLE I
Synthetic Seed Components

Component	Hydrated	Desiccated
High quality somatic embryo	Essential	Essential
Matured somatic embryo	Essential	Essential
Dormant or quiescent embryo	No	Essential
Protective coating	Beneficial	Beneficial
Germination arresting agents	Essential	No[a]
Germination inducer	Beneficial	No[a]
Inclusion of exogenous nutrients	No/essential[b]	No/essential[b]
Antimicrobial agents	No/essential[b]	No/essential[b]
Herbicides and insecticides	Beneficial	Beneficial
Desiccation protectants	No	Beneficial

[a]The desiccation process itself arrests germination, and germination occurs with the addition of water.

[b]Depends on whether the species is albuminous or exalbuminous.

(1) uncoated, desiccated somatic embryos, first described for orchardgrass (Gray *et al.*, 1985), (2) coated, desiccated embryos, as demonstrated with carrot (Kitto *et al.*, 1982), (3) encapsulated, hydrated embryos, using alfalfa (Redenbaugh *et al.*, 1984), and (4) hydrated embryos in a fluid-drilling gel, as suggested (but without success) with carrot (Drew, 1979). A fifth category could be composed of hydrated, uncoated somatic embryos. However, for mechanized delivery the latter would need to be either encapsulated or fluid drilled. Progress in developing these four types of synthetic seeds is discussed in this review. Currently, research is being conducted on synthetic seeds in a large number of laboratories around the world.

C. Development of Coatings

Citrus embryos were removed from seeds and coated with various compounds to develop a suitable synthetic seed coat (Kitto *et al.*, 1980). Of the eight compounds tested, polyoxyethylene was found to have the best film-forming ability, to readily redissolve in water, and to have a nondamaging effect on the embryos. Consequently, polyoxyethylene was chosen for further research on creating synthetic seeds using celery somatic embryos, although only multiple embryo synthetic seeds were produced (Kitto *et al.*, 1981, 1982, 1983, 1985a).

Fig. 1. Celery somatic embryos encapsulated in calcium alginate beads, approximately 5 mm in diameter.

Alfalfa and celery somatic embryos were used for developing hydrated synthetic seeds using various hydrogels, particularly sodium alginate (Fig. 1) (Redenbaugh *et al.*, 1986). Hydrated synthetic seeds were found to have a number of uses for propagation (Table II), although the technology is still in the research phase and has not yet been commercialized. Considerable advances have been made to date such that plant conversion was obtained from synthetic seeds of alfalfa, celery, and carrot planted in soil mix in an incubator. In addition, at Plant Genetics, Inc., we were successful in obtaining *in vitro* conversion for a number of other species (Table III).

TABLE II
Potential Benefits of an Artificial Seed Delivery System

Rapid propagation of desirable crop lines.
Genetic uniformity of plants.
Direct delivery of tissue-cultured plants to the field, thereby eliminating transplanting.
Deliver genetically unstable genotypes, such as transgenic plant lines that have meiotic instabilities.
Deliver sterile crop lines.
Reduce costs of vegetatively propagated elite lines.

[a]From Fujii *et al.*, 1987b.

TABLE III
Plants Produced from Synthetic Seeds Sown *In Vitro* and in Soil

	Species	Common Name
In Vitro	*Apium graveolens* L.	Celery
	Brassica sp.	Rapid-cycling brassica
	Daucus carota L.	Carrot
	Gossypium hirsutum L.	Cotton
	Lactuca sativa L.	Lettuce
	Medicago sativa L.	Alfalfa
	Oryza sativa L.	Rice
	Picea abies	Norway spruce
	Zea mays L.	Corn
In Soil	*Apium graveolens* L.	Celery
	Daucus carota L.	Carrot
	Medicago sativa L.	Alfalfa

II. UNCOATED, DESICCATED SYNTHETIC SEEDS

Increasingly, efforts to develop synthetic seed technology are shifting toward desiccated systems. All early work indicated that desiccation damaged the embryos, resulting in a lower conversion frequency. The first success in improving conversion via desiccation was for grape (Gray, 1987). Current progress in desiccation research is summarized in Table IV.

A. Alfalfa (*Medicago sativa*)

A collaborative group at the University of Guelph and Agriculture Canada is developing synthetic seed technology for alfalfa for mass propagation of self-sterile lines for hybrid seed production. Alfalfa somatic embryos have been dried to moisture contents between 8 and 15% and remained fully viable even after storage for 12 months at room temperature and humidity. Abscisic acid (ABA) was found to be essential for embryo conversion after desiccation. With ABA maturation, the conversion frequency reached 69% (60% average) for randomly picked somatic embryos and 90–100% for selected embryos. Without ABA, conversion of desiccated embryos was 0%. Conversion of nondesiccated, randomly picked embryos was 67%, which indicated that desiccation did not damage the embryos. Desiccated embryos were significantly more vigorous

TABLE IV

Research on Desiccated Somatic Embryos

Species	Extent of desiccation[a]	Desiccation duration	Conversion frequency[b]	Conversion of non-desiccated controls	Reference
Apium graveolens[c]	12% water content	3 days	86% survival	Not given	Janick *et al.*, 1989
Apium graveolens	Not given	36 hours	35%	Not given	Kaimori, 1988
Cary illinoensis	Not given	5 days	70% germination[d], a few plants produced	0%	Wetzstein *et al.*, 1989
Dactylis glomerata	13% water content	1 day	4–32%	Not given	Gray *et al.*, 1987
Daucus carota[c]	Not given	6.5 hours	3% survival	Not given	Janick *et al.*, 1989
Geranium	Not given	Not given	Not given	Not given	Roberts, 1990
Glycine max	Not given	Not given	30%	60%	Obendorf *et al.*, 1986
Glycine max	40–50% original volume	1–4 weeks	35–72% germination[d]	5% germination[d]	Hammatt *et al.*, 1987
Glycine max	Not given	1 week	26–100%[e] germination[d]	0–5% germination[d]	Parrott *et al.*, 1988
Glycine max	75% RH	4 or 7 days	79%	98%	Buchheim *et al.*, 1989
Medicago sativa	20% RH	20 days	33%	21%	Redenbaugh *et al.*, 1988a
Medicago sativa	8–12% water content	1–7 days	65–95%	65–95%[f]	McKersie *et al.*, 1989
Picea abies	Not given	Not given	90% germination[d]	Not given	Roberts *et al.*, 1989a
Triticum aestivum	10% water content	37 days	33% of callus formed embryos that germinated[d]	6% of callus formed embryos that germinated[d]	Carman, 1988a
Vitis longii	12% water content	21 days	34%	0%	Gray, 1989

[a]Somatic embryos desiccated at a relative humidity (RH) of 75% had a water content of 12%, at RH 85% the water content was 14%, and at RH 95%, 30% (Gray, 1989).

[b]Unless otherwise noted, the number given is percent conversion, which is production of complete plants containing roots, shoot, and leaves (or needles).

[c]Coated with polyoxyethylene before desiccation.

[d]Number given is percent germination, a term usually not defined in papers on somatic embryogeny, but used for various responses ranging from radicle elongation to complete plant formation.

[e]Frequency depended on genotype.

[f]Personal communication (B. McKersie).

than nondried ones, showing twice the shoot growth, with 45% increase in total fresh weight. However, all somatic embryos were considerably less vigorous than true seeds and appear to be less biochemically mature (Anandarajah *et al.*, 1989; Brown *et al.*, 1989; McKersie *et al.*, 1988, 1989; Senaratna *et al.*, 1987, 1989a, 1990).

B. Carrot (*Daucus carota*)

Carrot somatic embryos matured on ABA for 3 days were desiccated for 36 hours. When planted *in vitro*, up to 65% of the embryos survived and 35% formed plants (Kaimori, 1988).

C. Geranium (*Geranium* spp.)

Recently, a news report (Roberts, 1990) indicated that geranium somatic embryos were produced from lines being tested for resistance to bacterial blight. Although the embryos were desiccated, no data were given.

D. Grape (*Vitis longii*)

Desiccation of grape somatic embryos has been very successful in improving embryo germination and conversion. The water content of fully hydrated grape somatic embryos was 83%, whereas embryos dried under 75% relative humidity (RH) had a water content of 12%. Nondesiccated (fully hydrated) embryos had a germination of 5% or less and were considered dormant. In contrast, desiccation, even at 95% RH (water content of 30%), was sufficient to break dormancy and result in 34% conversion. This frequency was also achieved for embryos desiccated at 75% RH (Gray, 1988, 1989).

E. Orchardgrass (*Dactylis glomerata*)

Orchardgrass somatic embryos were desiccated to a water content of 13% for 24 hours, after which water equilibrium was reached. At this point, 32% of the desiccated embryos converted to whole plants. With longer periods of storage, the conversion frequencies decreased to 8% after 7 days and 4% after 21 days (Gray *et al.*, 1987). These results were typical for many species in which viability dropped quickly with storage of desiccated somatic embryos.

F. Soybeans (*Glycine max*)

The first reports on soybean somatic embryo desiccation presented little data but indicated 30% conversion after desiccation, which was significantly less than the 60% conversion for nondesiccated embryos. Additional information of the performance of these somatic embryos has not been reported (Obendorf *et al.*, 1986, 1988; Slawinska *et al.*, 1987).

Another laboratory has desiccated soybean embryos for 1 week in a sealed container. Germination was increased from 5% or less to 100% for some genotypes tested. Most interestingly, greenhouse-ready plants were produced from desiccated embryos in 2–3 months, compared with 9 months required for nondesiccated embryos (Parrott *et al.*, 1988).

G. Wheat (*Triticum aestivum*)

Wheat somatic embryos are difficult to separate from embryogenic callus and, as a result, germination frequencies are often given as a percent callus cultures with germinating embryos. In one experiment, significantly more desiccated callus cultures had somatic embryos that germinated (33%) compared with nondesiccated callus (6%). From the descriptions and photographs published, however, it would seem that no complete plants were produced (Carman, 1988a). The embryos were desiccated to 15% water content. Slow desiccation resulted in improved embryo viability, compared with fast desiccation (Carman, 1988b, 1989).

Oxygen levels and desiccation were tested for effects on callus growth and somatic embryo development and conversion. Lower oxygen improved embryogenic callus formation (but decreased overall callus growth), increased embryo production, decreased precocious germination, and improved embryo development and germination. ABA reduced precocious germination and reduced morphological abnormalities but did not increase germination. Using optimum conditions, germination frequencies up to 50% were achieved (Carman, 1988a).

III. COATED, DESICCATED SYNTHETIC SEEDS

A. Carrot (*D. carota*) and Celery (*Apium graveolens*)

Carrot and celery somatic embryos were coated with polyoxyethylene and desiccated. For carrot, only clumps containing multiple embryos, roots, and callus were coated, whereas individual celery embryos were isolated and coated. Survival up to 16 days was achieved for desiccated carrot embryos (Janick, 1988). Three percent of coated, desiccated carrot embryos survived, but no plants were produced (Kitto *et al.*, 1985a). Coated celery somatic embryos initially had 35% survival after desiccation (Kim *et al.*, 1987), but more recent work increased this to 86% with some of the coated embryos germinating and producing plants (Janick *et al.*, 1989). For both species, the embryos did not survive desiccation without the polyoxyethylene coating, except for celery embryos desiccated to 44% or greater water content (Kim *et al.*, 1989).

IV. COATED, HYDRATED SYNTHETIC SEEDS

The use of hydrated coatings was originally necessary because of poor desiccation tolerance of somatic embryos. Whereas desiccated embryos had poor conversion, hydrogel-encapsulated somatic embryos had conversion frequencies equal to noncoated embryos when planted *in vitro*. Only in the last few years, with significant advances in desiccation biology, has the emphasis shifted to desiccated synthetic seeds. However, considerable progress has been made with hydrated systems, and it is still too early to determine the ultimate applications. Certainly, approaches for recalcitrant crops may benefit from hydrated systems (Redenbaugh, 1990a).

A. Alfalfa (*M. sativa*)

Research on alfalfa hydrated synthetic seeds at Plant Genetics, Inc. has been summarized in various reports by Redenbaugh, Fujii, and Slade (see references). Additional data are presented later in this chapter.

A number of researchers in China are conducting research on synthetic seeds, including efforts for large-scale synchronous production of alfalfa somatic embryos (Anonymous, 1989a). Cold treatment of the embryos increased conversion from 52 to 57%. Desiccation of embryos for 5 days at 20°C resulted in 73% conversion. Encapsulation in sodium alginate reduced conversion to 62%, but if activated charcoal was added to the capsules, then conversion was 69%. The authors speculated that the charcoal improved respiration, perhaps by breaking up the alginate. Conversion experiments were also done in vermiculite in a greenhouse.

Only encapsulated embryos containing nutrient medium and sucrose produced plants (16% conversion) and the addition of charcoal improved this frequency to 23% (Zhong *et al.*, 1989).

B. Barley (*Hordeum vulgare*)

Several reports have appeared on barley synthetic seeds (Jain *et al.*, 1988; Datta *et al.*, 1989). The latter article described production of synthetic seeds using microspore-derived embryos. The embryos were produced directly without callus formation and encapsulated in a hydrated capsule (sodium alginate). The synthetic seeds germinated at 80%, producing a better root system than naked embryos that germinated at 62%. Encapsulated embryos were stored for 6 months without loss of germination, whereas naked embryos died after 2 weeks. This latter result is surprising since all other reports on embryo encapsulation in alginate reported either a neutral or negative effect of the gel on embryo germination or viability. In some experiments each barley microspore produced one embryo, but overall the number of anthers producing embryogenic microspores was 660 out of a total of 3,000. This gave 3,680 embryos, of which 80% matured and were plated on modified MS (Murashige and Skoog, 1962) medium. Of these, 860 produced complete plants, giving a conversion frequency of 29%. Of 66 plants transferred to the greenhouse, 53 produced fully fertile plants. From 3,000 embryogenic microspores, 53 fertile plants were produced (Datta *et al.*, 1989).

C. Carrot (*D. carota*) and Celery (*A. graveolens*)

Hybrid celery can be produced only through laborious hand pollination and, as a result, has not been commercialized. Vegetative propagation has been proposed for hybrid celery but not used because of high cost. Because celery somatic embryos are easily produced, considerable research has been conducted to develop a low-cost propagation system. In one report, celery somatic embryos were encapsulated in hydrogels to form synthetic seeds, but no details on the hydrogel or conversion of the embryos were presented. However, there was significant variation among celery tissue culture lines, particularly in the numbers of embryos produced and the percent conversion (Lutz *et al.*, 1985).

Other work on celery synthetic seeds has shown a need for an artificial endosperm and an outer coating to control desiccation and facilitate

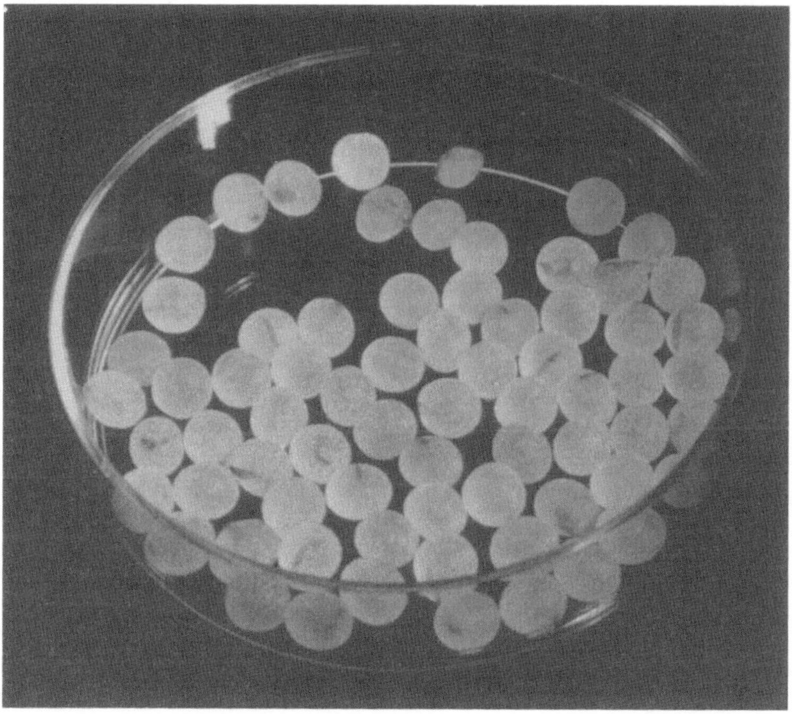

Fig. 2. Celery somatic embryos coated with Elvax 4260 polymer to retard desiccation and reduce tackiness (Redenbaugh *et al.*, 1987a).

handling (Fig. 2) (Redenbaugh *et al.*, 1987a). For celery in our laboratory, soil conversion was affected by encapsulation and the addition of sucrose. Soil conversion is defined as germination and production of complete plants in soil or greenhouse mix under nonaseptic incubator, greenhouse, or field conditions. For celery synthetic seeds, we achieved repeatable conversion in soil in a growth chamber only when sucrose and an antimicrobial solution were included in the hydrogel capsule (Fig. 3). Conversion of naked embryos in soil was comparable to encapsulated embryos with sucrose and an antimicrobial solution, indicating the somewhat detrimental effect of calcium alginate. The antimicrobial solution was essential for controlling contamination of sucrose. Sucrose loading of somatic embryos during maturation was also effective in improving the performance of the embryos under soil conversion. A 1.5-fold Schenk and Hildebrandt (SH) nutrient solution (Schenk *et al.*, 1972) with either 4.5 or 6% sucrose was most effective (Fig. 4).

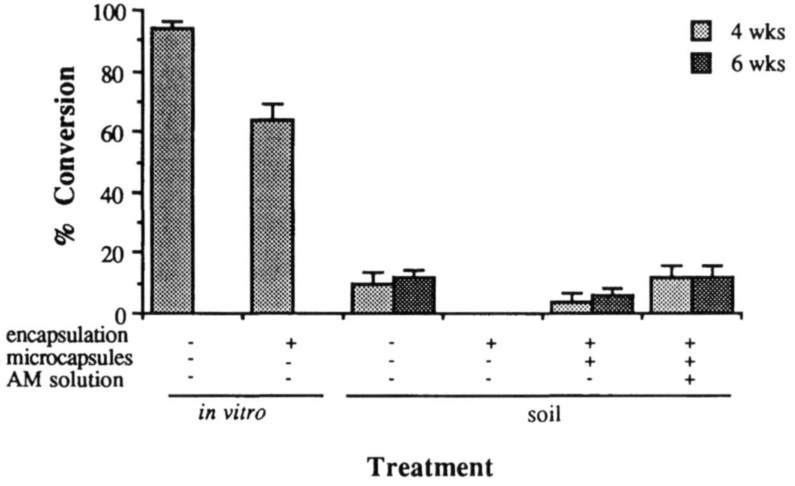

Fig. 3. Effect of encapsulation and addition of microcapsules and antimicrobial solution on conversion of celery somatic embryos. The antimicrobial (AM) solution consists of a number of fungicides, bacteriostats, and antibiotics. Conversion measured 4 and 6 weeks after planting. Bars equal one standard error. Treatments were applied (+) or not applied (−) to the embryos.

Another report found that carrot somatic embryos, matured with ABA and encapsulated in sodium alginate, germinated *in vitro* at 45% frequency. Subsequently, plants were transferred to the greenhouse and taken to harvest, with no apparent abnormalities (Jain, 1987).

High quality carrot somatic embryos were produced using novel in-

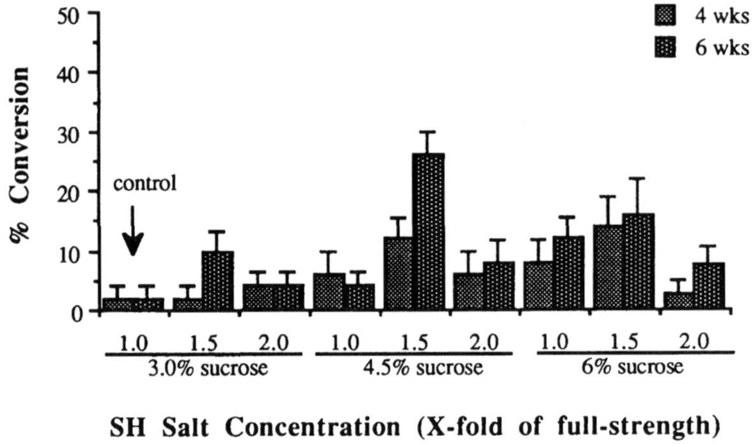

Fig. 4. Effect of SH and sucrose concentration during maturation on soil conversion of celery. Conversion measured 4 and 6 weeks after planting. Bars equal one standard error.

duction techniques. Large embryos were induced using high sucrose, cadmium, or sodium hypochlorite, then encapsulated in sodium alginate and converted *in vitro* at 30–50%. When the standard induction agent, 2,4-dichlorophenoxyacetic acid (2,4-D) was used, embryo conversion was only 15%. The synthetic seeds were germinated in an empty petri dish and then transplanted to vermiculite in an incubator (Kamada *et al.*, 1988, 1989).

In China, researchers have encapsulated carrot somatic embryos in sodium alginate with half-strength MS medium (including sucrose) and sown the synthetic seeds directly in vermiculite and soil under sterile conditions. Germination of 70% was obtained, while *in vitro* germination was 93%. Although plants were transferred to the greenhouse, actual conversion frequencies were not given. Encapsulated carrot embryos could be stored for 1 month without decrease in germination frequency (Li, 1987; Li *et al.*, 1989).

D. Lettuce (*Lactuca sativa*)

Lettuce F_1 adventitious buds were encapsulated in sodium alginate. Shoot and rooting frequencies were not given, but complete plants were produced (Hayashi *et al.*, 1988a,b). To improve rooting and shoot formation from alginate capsules, 2,4-D levels were altered and the buds were partially dried, resulting in *in vitro* rooting between 60 and 80% (Onishi *et al.*, 1988a,b). More than 20,000 encapsulated and naked lettuce buds and celery embryos were sown directly in the greenhouse (Anonymous, 1988a). Other research efforts at Kirin Brewery Co., Ltd. are directed toward rice synthetic seeds. The development of rice synthetic seeds for hybrid propagation has also been proposed in France (Courtois, 1988).

E. Tomato (*Lycopersicon esculentum*)

Research on synthetic seeds in France began in 1985 at the Laboratoire d'Amelioration des Plantes at Orsay and by Francereco. Other research began in 1986 under a consortium called Eureka by Limagrain, Rhone-Poulenc Agrochimie and Francereco (Nestle) with focus on tomato synthetic seeds. Success has been limited, however, and no reports of somatic embryogenesis in tomato have appeared. The Eureka group has proposed that in carrot up to 80,000 embryos per liter per day could be produced (Nouaille *et al.*, 1988).

F. Trees

Considerable progress has been made in somatic embryogeny of trees for forest, fruit, ornamental, and specialty use trees (Redenbaugh *et al.*, 1989). Although somatic embryos from several conifer species, such as Norway spruce, loblolly pine, and sugar pine (Gupta *et al.*, 1987; Gupta, 1988), have been encapsulated in sodium alginate, germination and conversion have been poor.

Bapat (Bapat *et al.*, 1987) reported encapsulating *Morus indica* (mulberry) shoot buds in sodium alginate beads and in agar blocks, with alginate giving a better response in terms of plant production. Encapsulated shoot buds could be stored at 4°C for 45 days without loss of viability. Bud "germination" was 50–90% *in vitro* and 10% on agar water or on moistened filter paper. Buds sown directly in soil failed to develop further. Agar was also tested for encapsulation but was inferior to alginate as measured by bud "germination."

Somatic embryos of *Santalum album* (sandalwood) were encapsulated in sodium alginate. The synthetic seeds could be stored at 4°C for 45 days. Overall germination was poor, at less than 10%. No nonencapsulated embryos were tested, nor was germination obtained in soil (Bapat *et al.*, 1988, 1989).

In one interesting example, excised radiata pine zygotic embryos were placed cotyledon first into a small amount of agar medium contained within an aluminum cup such that only the root stuck out. These "artificial seeds" were germinated with growth rates slightly less than intact seeds. The authors suggested that this process could be useful for somatic embryos (Teasdale *et al.*, 1986).

G. Other Species

Demarly (personal communication, 1987) indicated that his laboratory had encapsulated somatic embryos of alfalfa, carrot, date palm, grapevine, and soybean in sodium alginate but did not indicate whether any of the synthetic seeds germinated. Another report indicated that alginate-encapsulated date palm somatic embryos germinated, but no frequencies were given (Daikh *et al.*, 1987).

Other research is being conducted on synthetic seeds of recalcitrant species, to better understand their development and physiology, particularly to improve germ plasm storage. Several groups are active with recalcitrant crops, including International Board for Plant Genetic Resources with D. Gray and R. Litz (University of Florida), D. Come (Physiology Laboratory, Meudon), and Limagrain (Nouaille *et al.*, 1988).

V. UNCOATED, HYDRATED SYNTHETIC SEEDS (FLUID DRILLING)

The use of fluid drilling for delivery of somatic embryos was first discussed for carrot (Drew, 1979), but only three plants were produced from embryos sown on a sugar-free basal medium. Lawrence (1981) outlined the requirements for large-scale propagation using somatic embryogenesis and fluid drilling for delivery and suggested target crops, including tomato, pepper, melons, lettuce and celery. Attempts to sow carrot somatic embryos under greenhouse conditions using fluid drilling resulted in 4% embryo survival but no germination or plant production (Baker, 1985).

Sweet potato (*Ipomoea batatas*) somatic embryos were sown in fluid drilling gels supplemented with various carbohydrates. Plant formation occurred for 20–25% of the embryos using N-gel (Natrosol) with maltose, sucrose, and glucose. Other gels resulted in reduced or complete inhibition of embryo germination. Sucrose alone was not an adequate carbohydrate source for plant formation (Cantliffe *et al.*, 1986; Schultheis *et al.*, 1986, 1988).

VI. MATURATION OF SOMATIC EMBRYOS

Application of synthetic seed technology is limited by the quality of somatic embryos. Because our knowledge of somatic embryogeny is insufficient, much of the synthetic seed research is focused on maturation of somatic embryos. To better decipher somatic embryogeny, we developed highly efficient protocols for alfalfa in which the processes were compartmentalized to allow for testing of specific parameters (Fig. 5). In this review, we will focus only on a few of the steps, although all are critical and poor results occur if not followed explicitly.

A. Pretreatment

Yields of alfalfa somatic embryos were significantly improved when amino acids such as proline, alanine, arginine, or glutamine were added to the callus maintenance medium. Up to 100 times more embryos were produced (Stuart *et al.*, 1990; Slade *et al.*, 1990). Organic acids applied during the last subculture of callus growth ("pretreatment") were found to increase the numbers of embryos formed, the quality (conversion) of

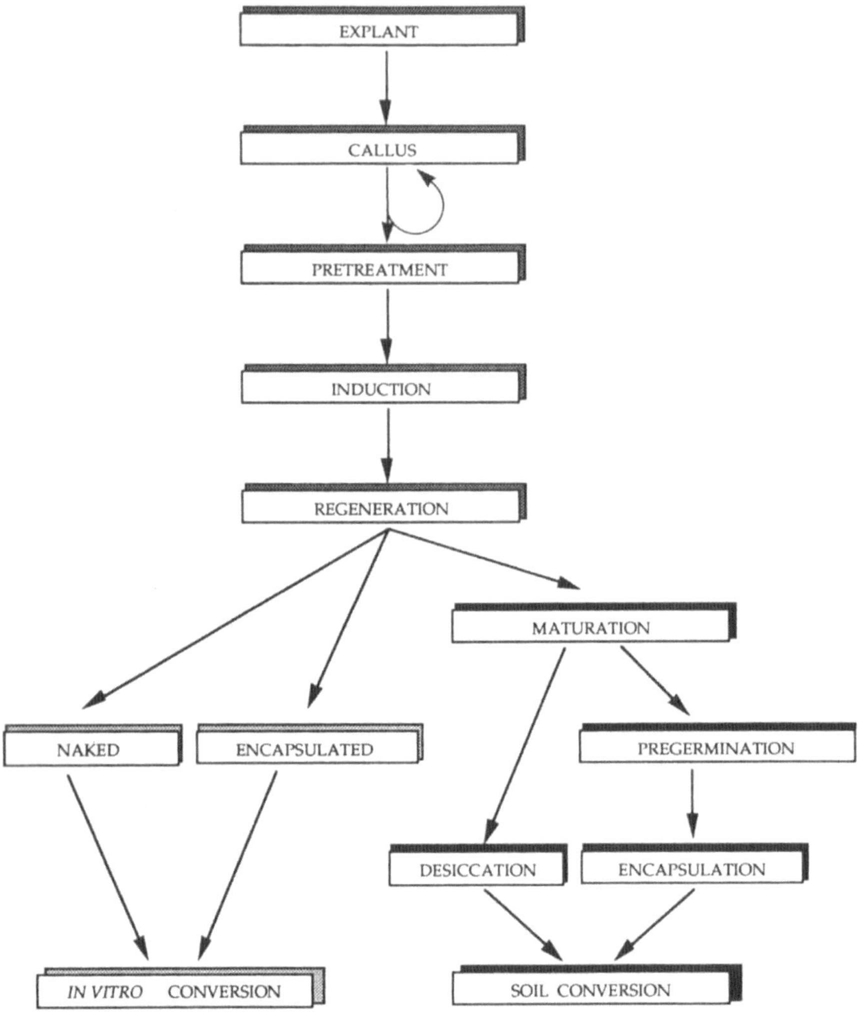

Fig. 5. Specific steps for alfalfa somatic embryo formation and conversion to plants (Slade *et al.*, 1989). Callus is produced from petiole explants and bulked up on 25 ml SH (Schenk *et al.*, 1972), 25 μM naphthaleneacetic acid (NAA), and 10 μM kinetin (SHNAAK medium) (Walker *et al.*, 1981). During the pretreatment stage, the callus is grown 3 weeks on a citrate medium consisting of 25 ml SH (without KNO_3), 25 μM NAA, 10 μM kinetin, 20 mM K-citrate, and 25 mM glutamine (Stuart *et al.*, 1990; Slade *et al.*, 1990). Callus is then placed on induction medium, 25 ml SH, 50 μM 2,4-D, and 5 μM kinetin (SHIM) (Walker *et al.*, 1981). For regeneration of somatic embryos, a hormone-free medium is used: 10 ml SH, 10 mM $NH_4H_2PO_4$, and 50 mM L-proline (Regen) (Stuart *et al.*, 1984a,b). At this point somatic embryos are produced. In early work, the embryos were then planted *in vitro* on 25 ml 1/2 SH (no sucrose), 1.5% maltose, and 25 μM gibberellic acid (GA_3), either naked or encapsulated (left side of diagram) (Redenbaugh *et al.*, 1986). More recently, the somatic embryos were matured on SH with 5 μM ABA (SHABA) for 3 weeks (right side of diagram) (Fujii *et al.*, 1989). The embryos are then either desiccated and planted directly in potting mix in a growth chamber or pregerminated for 2 days and encapsulated in sodium alginate followed by soil conversion. Soil conversion is now the current protocol for embryo evaluation (Fujii *et al.*, 1989).

TABLE V

Effect of Organic Acids in Callus Medium (Pretreatment Stage)

Organic acid type and concentration	Embryos per Plate		% Conversion	
	Ave.	Std. error	Ave.	Std. error
SHNAAK[a] (Control)	242	10	32.0	2.0
SHNAAK + 20 m*M* K-citrate	323	12	47.0	2.7
SHNAAK + 60 m*M* K-malate	309	10	49.6	3.7
SHNAAK + 70 m*M* K-tartarate	351	13	48.4	4.7

Source: Stuart *et al.*, 1990; Slade *et al.*, 1990.

[a]SHNAAK = SH medium (Schenk *et al.*, 1972), 25 μM naphthaleneactic acid (NAA), and 10 μM kinetin (Walker *et al.*, 1981).

the embryos (Table V), and the amount of seed-specific storage proteins (Table VI) (Stuart *et al.*, 1990). Calculations were done (Table VII) to determine the most efficient through-put system. Although the organic acid pretreatment significantly reduced callus growth during the last subculture, the total number of plants produced was almost double the control treatment. The pretreatment effect occurs before embryo formation (before induction on 2,4-D) and may be selecting for more embryogenic cells or altering cellular metabolism toward cell differentiation and away from cell proliferation.

B. Maturation

Embryo maturation with abscisic acid (ABA) was essential for obtaining high soil conversion. ABA-matured alfalfa somatic embryos had an average of 50% soil conversion when planted in an incubator (highest frequency to date was 90%) and up to 64% soil conversion in a high humidity greenhouse. Optimum concentration of ABA for maturation of alfalfa embryos was 5 μM.

TABLE VI

Effect of Organic Acids during Pretreatment Stage on Seed Storage Protein

Treatment	Micrograms protein per gram somatic embryo	Percent conversion	
		Ave.	Std. error
SHNAAK[a] (Control)	135	24	6.1
SHNAAK + 20 m*M* K-citrate	252	40	3.6

[a]SHNAAK = SH medium (Schenk *et al.*, 1972), 25 μM NAA, and 10 μM kinetin (Walker *et al.*, 1981).

TABLE VII
Alfalfa Mass Balance.[a] Efficiency of Plant Production from 1 g Callus

	SHNAAK[b] (control)	K-Malate	K-Citrate
Maintenance			
From 1 g callus	3.9 g	1.6 g	2.2 g
Regeneration			
Embryos per plate	143	227	232
Embryos from 1 g callus	8,500	5,600	8,000
Conversion			
% Conversion	21	39	43
Plants from 1 g callus	1,800	2,200	3,400

[a] Slade *et al.*, 1990
[b] SHNAAK = SH medium (Schenk *et al.*, 1972), 25 μM NAA, and 10 μM kinetin (Walker *et al.*, 1981).

ABA was also critical for soil conversion of carrot (Fig. 6) and celery somatic embryos, with 1–5 μM ABA giving the highest conversion frequencies. Even higher concentrations of ABA were beneficial for soil conversion for celery (Fig. 7).

The effect of ABA was not osmotic but due to accumulation of embryo-specific storage reserves in alfalfa somatic embryos, such as carbohydrates (Fujii *et al.*, 1988, 1989). In contrast, water relations were

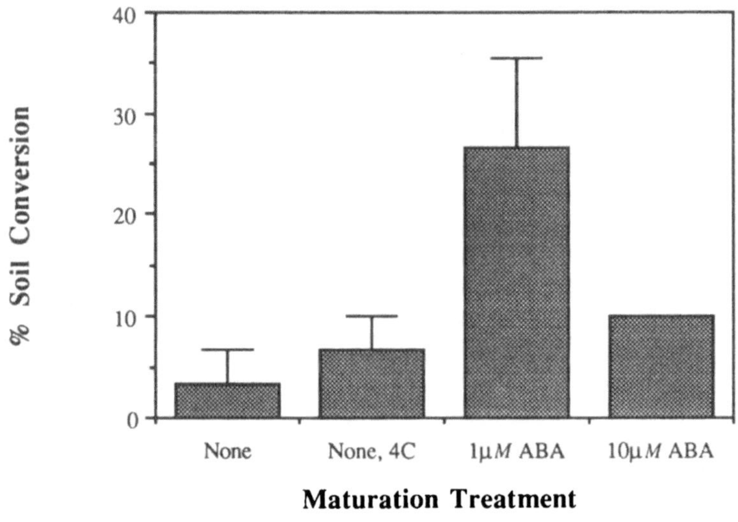

Fig. 6. Effect of ABA maturation on carrot somatic embryo conversion. Maturation treatments were (1) no ABA or cold maturation, (2) no ABA but the embryos were matured 21 days at 4 C, (3) 1 μM ABA for 21 days, or (4) 10 μM ABA for 21 days. Bars equal one standard error.

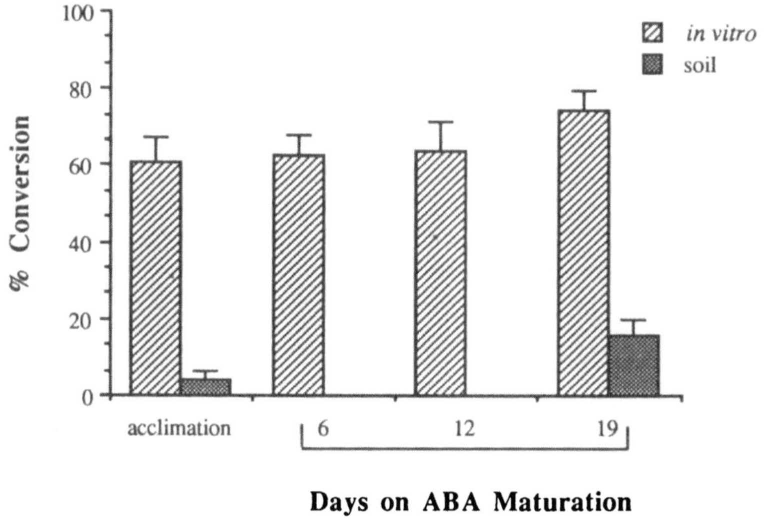

Fig. 7. Effect of acclimation treatment (for 4 days) or SH + 50 μM ABA maturation on celery somatic embryo conversion *in vitro* or in soil. Bars equal one standard error.

proposed as controlling maturation and germination in mustard seed development (Fischer *et al.*, 1988).

The interaction of ABA and desiccation tolerance has been studied with several species, beginning with carrot (Kitto *et al.*, 1985b). ABA was also a critical component of the maturation medium for obtaining survival after desiccation for celery somatic embryos (Park *et al.*, 1988) and for producing "mature" white spruce somatic embryos (Roberts *et al.*, 1989b).

Desiccation of soybean somatic embryos has been studied in several laboratories. One group desiccated secondary embryos produced from zygotic embryos and recovered complete plants. The average plant production was quite low with only two secondary embryos per explant. However, 35% of the desiccated embryos germinated compared with less than 5% for nondesiccated embryos (Hammatt *et al.*, 1987). In another report, desiccated somatic embryos germinated between 26 and 100% (depending on genotype), compared with 0–5% germination without desiccation (Parrott *et al.*, 1988). In another laboratory, soybean conversion frequency was increased to 96% for one genotype when the somatic embryos were first matured on a 10% sucrose maturation medium. These embryos, however, did not show improvement in conversion frequency or rate after subsequent desiccation (Buchheim *et al.*, 1989). The length of time that somatic embryos were on the 10% sucrose maturation medium was also important, with higher conversion frequencies after

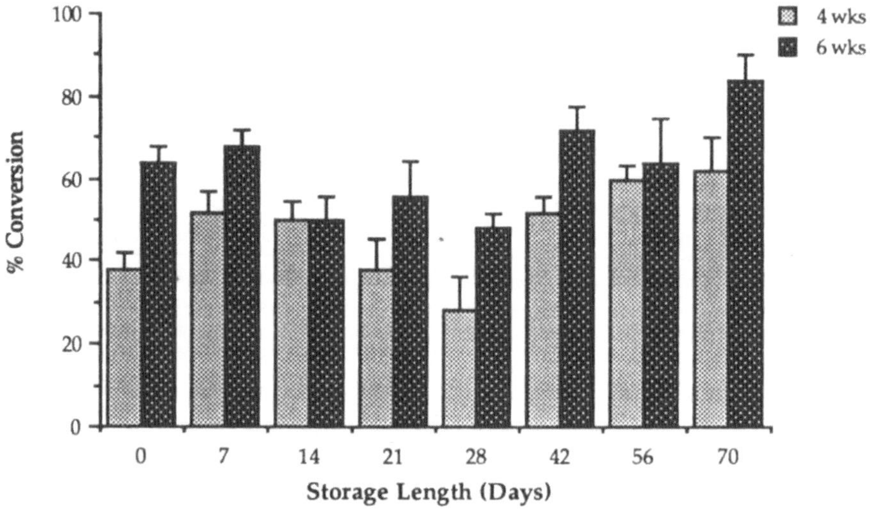

Fig. 8. Long-term cold storage (4C) of naked alfalfa somatic embryos. Conversion scored after 4 or 6 weeks. Bars equal one standard error.

45 days vs. 30 days (Shoemaker *et al.*, 1988). Embryo desiccation may be involved in reduction of endogenous ABA levels (Hammatt *et al.*, 1987).

The overall effect of ABA is quite pronounced in a number of species, with clear improvements in embryo quality as measured by efficacy in soil conversion and conversion after desiccation. The precise mechanism of ABA maturation still needs to be resolved.

C. Storage

After ABA maturation hydrated, nonencapsulated alfalfa somatic embryos could be stored at 4°C for over 2 months without loss of conversion ability (Fig. 8). After approximately 6 months' storage, the quality of embryos begins to decline. Storage is a critical factor for producing large quantities of synthetic seeds for commercialization.

D. Conversion

Greenhouse conversion of somatic embryos was dependent on two major factors: embryo quality (as already discussed) and the watering system. A narrow range of watering conditions was essential. A heavy misting system caused embryo browning and rotting, whereas soil drying caused embryo desiccation and death (Fujii *et al.*, 1986, 1987b). This sensitivity of somatic embryos to greenhouse watering was much more severe than for true seeds. Whether the use of desiccated somatic embryos will eliminate such watering sensitivity has not yet been tested.

VII. PROBLEMS WITH SYNTHETIC SEEDS

A. Respiration in Hydrogel Beads

In previous papers on hydrated, encapsulated somatic embryos, gas exchange was found to be poor in alginate gels, resulting in lower conversion frequencies compared with nonencapsulated embryos (Redenbaugh *et al.*, 1988b, 1990a). In approximately half of our experiments, encapsulation resulted in a significant decrease in conversion, as in Fig. 9. At other times, there were no significant differences nor was the percent alginate critical (Fig. 10). Other reports confirm this decline in conversion of somatic embryos encapsulated in sodium alginate (Zhong *et al.*, 1989), although one report showed that encapsulation improved conversion (Datta *et al.*, 1989). In a recent report, steep oxygen gradients were found in single calcium alginate beads containing *Enterobacter cloacae*, with up to 70% of the bead being free of oxygen at the end of the logarithmic growth phase (Beunink *et al.*, 1989). One potential solution to the respiration problem is to provide an oxygen-enriched environment within the gel.

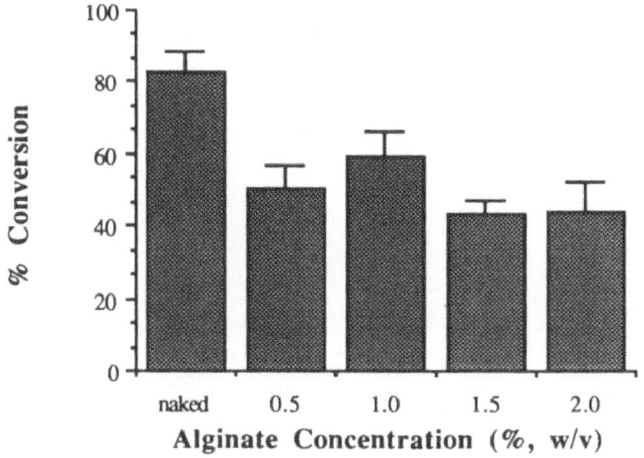

Fig. 9. Effect of encapsulation on *in vitro* conversion of celery somatic embryos. Plants were scored for conversion four weeks after planting. Bars equal one standard error.

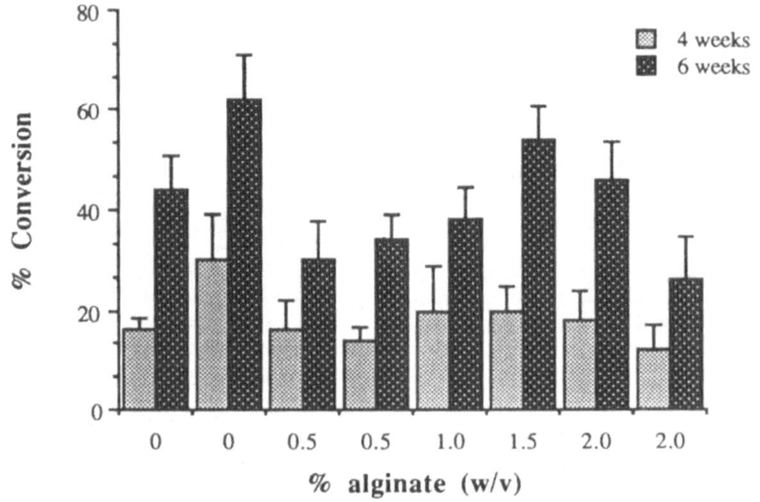

Fig. 10. Effect of alginate concentration on soil conversion of alfalfa somatic embryos. Plants were scored for conversion at four and six weeks after planting. Bars equal one standard error.

Perfluorochemical (PFC) oils have been used as organic oxygen carriers in a variety of conditions where gas transport is limiting. For growth of microbial cells, addition of PFCs to the culture medium resulted in a variety of responses such as increased oxygen transfer, enhancement of culture growth and productivity, and increased superoxide dismutase activity. In mammalian cell cultures, PFCs caused increased cytoplasmic vacuolation, reduced adherence, and other responses (King *et al.*, 1989). No reports have yet appeared on the use of PFCs with plant cell cultures. PFCs could be emulsified with the sodium alginate gel before complexing with calcium to provide encapsulated somatic embryos with a readily available source of oxygen for respiration.

B. New Coatings

Although most research on synthetic seeds has focused on two polymers, sodium alginate and polyoxyethylene, other polymers should be tested, both existing and newly developed ones. Handling of hydrated somatic embryos is difficult on a large scale because of tackiness. Except under fluid conditions, the embryos do not flow. To overcome these limitations, we tested several fumed silica compounds such as Tullanox®, which is very hydrophobic, and Cab-O-Sil® (both from Tulco, Inc.), which is slightly hydrophilic. These powder coatings were ap-

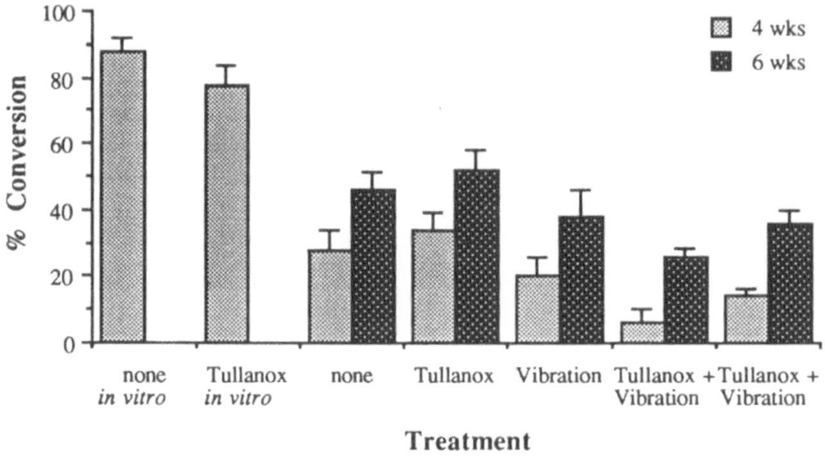

Fig. 11. Effect of Tullanox 500 and vibration singulation on conversion of alfalfa somatic embryos in soil. Conversion was scored 4 and 6 weeks after planting. Bars equal one standard error.

plied by rolling or mixing the embryos in the powder. In experiments with alfalfa, a Tullanox coating had either no or a slightly negative effect on soil conversion compared with no coating. However, when the embryos were subjected to the dehydrating conditions of a vibratory bowl (such as that used by the UCD encapsulation machine) with and without Tullanox coating, the conversion frequency decreased (Fig. 11). A similar decrease in conversion was observed when the coated embryos were encapsulated either by hand using a dropping system or via machine (Fig. 12). The Tullanox coating was also used to coat alginate-

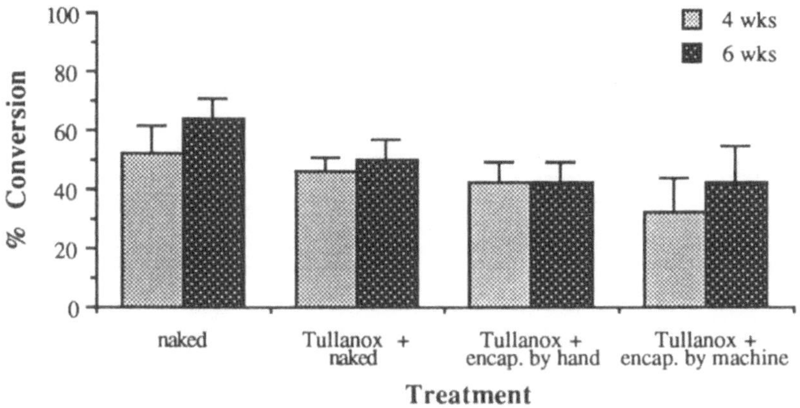

Fig. 12. Effect of Tullanox and encapsulation (by hand or machine) on alfalfa conversion in soil. Conversion was scored 4 and 6 weeks after planting. Bars equal one standard error.

encapsulated somatic embryos to eliminate capsule tackiness, which enabled the capsules to be handled with a belt seeder. Cab-O-Sil was similar to Tullanox in providing a powder coating, but because it is hydrophilic, it absorbed water over several hours, resulting in tacky embryos.

Researchers at Iowa State University are examining water-regulating polymers for seed coating, including ones that were originally used for tropical seeds (Heacox, 1989). Such polymers that regulate uptake of water have application for coating desiccated somatic embryos. Rehydration of embryos is a critical component for germination and conversion. Damage has been reported when dried soybean somatic embryos were germinated too quickly (Obendorf *et al.*, 1988; Obendorf, personal communication). A slow rate of water uptake during rehydration appeared to improve germination frequency (Senaratna *et al.*, 1989b). Alfalfa seed coats, testa, and endosperm (with decreasing effect) inhibit germination of the zygotic embryos and may function, in part, to facilitate embryo maturation (Xu *et al.*, 1989). Appropriate polymers may also provide protection against fungi (Heacox, 1989), eliminating the need for fungicide treatments of the coated embryos. This would be particularly beneficial for albuminous species for which contamination occurs with the addition of a synthetic endosperm.

Silicone coatings have also been proposed for synthetic seeds, primarily because of their antifungal properties and permeability to water vapor and oxygen. Potential coatings are polydiorganosiloxanes such as dimethylsiloxanes, methylphenylpolysiloxanes, methylvinylpolysiloxanes, dimethylsiloxane, and methylvinylsiloxanes. These coatings were applied by dunking seeds into the water-soluble silicone latexes and then air-drying overnight. The resultant film was a soft rubbery coating. However, no somatic embryos were coated using this process (Ryan *et al.*, 1988).

C. Creating a Bio-Coating Seed Coat

A speculative, alternative research direction for coating somatic embryos for synthetic seed formation is to redirect cell differentiation so that specific tissue types are induced, such as endosperm and testa (including cuticle and subcuticle cells). With precise control of cell differentiation, these tissues could be induced on the outer surface of somatic embryos or induced from other cells or tissues. In the latter case, once the endosperm and testa tissues were formed they would then be used to coat somatic embryos. Certainly, such approaches are beyond the cellular control we currently can exert but may be possible in the future.

VIII. APPLICATIONS

Synthetic seed technology can offer a low-cost clonal propagation system for a variety of species, such as hybrid rice, potato, geraniums, European seedless cucumber, garlic, and Gerbera daisy (Fujii *et al.*, 1987b).

A. Alfalfa

Potential use of synthetic seed technology for alfalfa is not for propagation but rather as a tool for breeding and variety development, such as for hybrid alfalfa. For this to occur, however, direct planting into at least a greenhouse and preferably into the field is required. Greenhouse sowing has been accomplished for alfalfa with conversion frequencies equal to that obtained in growth chambers (Fujii *et al.*, 1989). During the past several years, synthetic seeds were sown directly into the field in work conducted by Plant Genetics' researchers. Although the conversion frequency has been low to date (Fig. 13), this success with field planting supports the concept of breeding hybrid alfalfa and developing varieties with improved heterosis (Redenbaugh *et al.*, 1990b).

Fig. 13. Direct sowing of alfalfa synthetic seeds in field in Yolo County, California. Immediate watering was essential to achieve field conversion. Plants from somatic embryos were found to grow more slowly than seedlings.

B. Forestry

Forestry provides one of the most valuable opportunities for synthetic seed technology because of the need for vegetative propagation and the lack of other cost-effective methods. Recently, thousands of complete trees were produced from several conifer species, including Eastern white pine, white spruce, black spruce, red spruce, Norway spruce, Japanese larch, European larch, and hybrid larch (W. Cheliak, personal communication, 1990). The use of synthetic seeds in forestry has the potential to reduce the tree breeding cycle 5–20 years by eliminating the need for seed production orchards. Cryopreservation of embryogenic tissue will be a critical component for applying synthetic seed technology to forestry (Redenbaugh *et al.*, 1989).

Other organizations are also active in synthetic seed development of tree crops. Weyerhaeuser Company is conducting research on conifer synthetic seeds, including attempts to improve the quality of somatic embryos (Anonymous, 1989b). The New Zealand Ministry of Forestry and Elders Resources NZFP formed a partnership for producing *Pinus radiata* (radiata pine) synthetic seeds. The current focus is on embryo production in bioreactors. However, no data have been published (Anonymous, 1988b). Another company, B.C. Research, is developing desiccated spruce synthetic seeds (B. Sutton, B.C. Research, personal communication, 1990).

C. Lettuce and Celery

Kirin and Plant Genetics conducted field trials in Japan with 20,000 F_1 synthetic seeds in 1988 with celery somatic embryos and lettuce adventitious shoot buds. Kirin is also working on synthetic seeds for rice and flower crops. Their objective is "to develop synthetic seeds that can be sowed directly to fields and to produce synthetic seeds from superior plants" (Kirin biotechnology report, 1990).

D. Other Applications

Hydrated encapsulation of somatic embryos from recalcitrant species may be useful for clonal propagation and provide a method for increasing the very short storage life of the seeds (Redenbaugh, 1990a). Other applications of synthetic seed technology for agronomic crops have been recently reviewed. Except for sugarcane and alfalfa, most agronomic crops will benefit little from this technology (Redenbaugh, 1990b).

IX. MECHANIZATION

A. Bioreactor Development

For horticultural crops, tens of thousands to millions of somatic embryos will be needed for large-scale commercialization. For agronomic crops, millions to billions of embryos are required. Current embryo production methods clearly are inadequate for such numbers. All bioreactor development to date for embryo production has used standard systems such as impeller-driven or air-lift vessels.

One report examined 10 different types of bioreactors and found a spin filter device to give the best results with carrot, because the used medium could easily be removed. However, the authors did not do comparative studies with other bioreactors (Styer, 1985).

Another paper determined that a vibrating bioreactor using a hanging stir bar was the best system for alfalfa somatic embryo production (Virtis Model BSF 2000). Control of dissolved oxygen was critical for this work. High shear force bioreactors were unsuitable for embryo production (Chen *et al.*, 1987), a result seen by others (Stuart *et al.*, 1987). The number of embryos per milliliter of suspension culture was reduced in bioreactors (9 embryos/ml) compared with flasks on a shaker (44 embryos/ml). The quality of embryos produced in the bioreactor was equal to those in suspension culture, with 30–80% of the embryos germinating, depending on embryo size. However, no comparison was made with embryos produced on agar medium, nor were conversion frequencies given (Chen *et al.*, 1987).

In our laboratory alfalfa and celery somatic embryos produced on agar-based plates were always significantly higher quality based on conversion frequencies than any flask- or bioreactor-produced embryos (Stuart *et al.*, 1987). Further research is needed to identify critical parameters for high quality embryo production on a large scale. Recently, one report discussed kinetic growth and substrate utilization for alfalfa cell growth. Although the intent was not for embryo production, a two-phase cell growth dynamic was measured (McDonald *et al.*, 1989). We anticipate that to optimize embryo production in bioreactors, a multiphasic dynamic system will be necessary.

B. Encapsulation Mechanization

Research on mechanization of alginate bead handling has focused on field delivery. Gautz *et al.* (1987a,b) were able to separate hydrogel capsules using a vibrating box. Capsules that were uncoated did not separate well, whereas wax-coated capsules could be separated using vibration (Fig. 2). Although not identified in the article, the wax coating was that previously described (Redenbaugh *et al.*, 1987a).

Mechanical planting of hydrated capsules (without seeds or embryos) was done using vacuum seeders, such as a Gaspardo, or belt seeders, such as the Stanhay. These planters were retrofitted to accurately meter the capsules (Upadhyaya *et al.*, 1987).

During one field trial conducted at Plant Genetics using alfalfa synthetic seeds, a Stanhay planter was used for sowing. Although the capsules were properly planted, the quality of the somatic embryos was poor and no plants were produced (Fujii *et al.*, unpublished data).

Recently, a sophisticated encapsulation machine was constructed by the University of California, Agricultural Engineering Department, in conjunction with Plant Genetics, Inc., for the purpose of encapsulating seeds and somatic embryos in calcium alginate beads (Garrett *et al.*, 1989). The machine consists of six units connected to a central computer that records the number of filled capsules and blanks and prepares data for printing labels for individual bags of synthetic seeds. The machine is designed such that one person can operate all six units. Individual units can also be operated independently. A schematic drawing of the machine is shown in Fig. 14. Each unit of the machine consists of a dual ramp vibratory bowl to move the somatic embryos to one of two nozzles (Fig. 15). The movement of alginate into the nozzles is controlled by a positive displacement, adjustable stroke, air pressure-driven pump. Seed singulation on the vibratory bowl is controlled by a panel box on each unit (Fig. 16). Other functions are controlled by a microprocessor-based system which independently coordinates the two separate capsule-producing and inspecting channels.

The process of encapsulation using the six-unit machine is described in Fig. 17. Each nozzle is calibrated to produce approximately one encapsulated seed or embryo per second. With the current nozzle assembly, the size limitation of encapsulatable seeds or embryos is approximately 2–6 mm. Smaller seeds and embryos are not well singulated, whereas larger ones clog the nozzle tube. The machine is well suited for encapsulation of seeds and embryos similar to the size of tomato seeds and has been used primarily for this purpose (Garrett *et al.*, 1990).

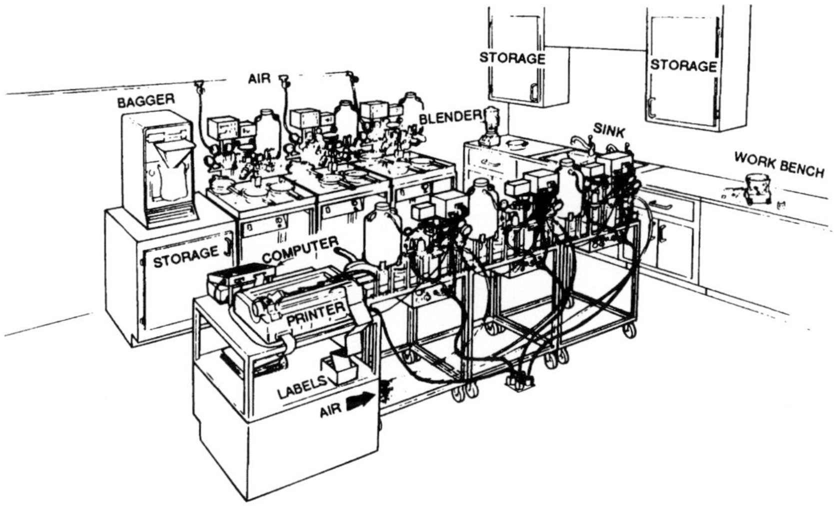

Fig. 14. Encapsulation machine showing six individual units, each with a dual nozzle, a bagger to package each lot of synthetic seeds, a computer to record the efficiency of each unit and to control the printer for preparing labels for each lot (Garrett *et al.*, 1989).

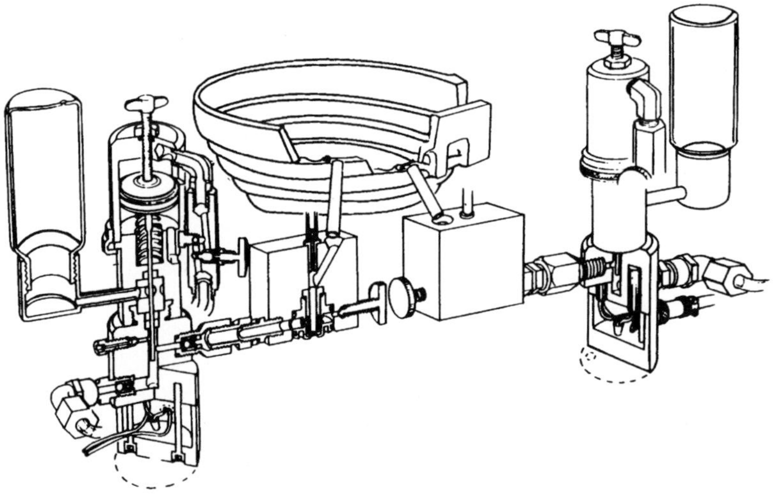

Fig. 15. Vibratory bowl and dual coaxial tube nozzle assembly of one unit of the encapsulation machine. Embryos move up the bowl and into one of the two nozzles. Sodium alginate is pumped into each nozzle for encapsulating the embryos (Garrett *et al.*, 1989).

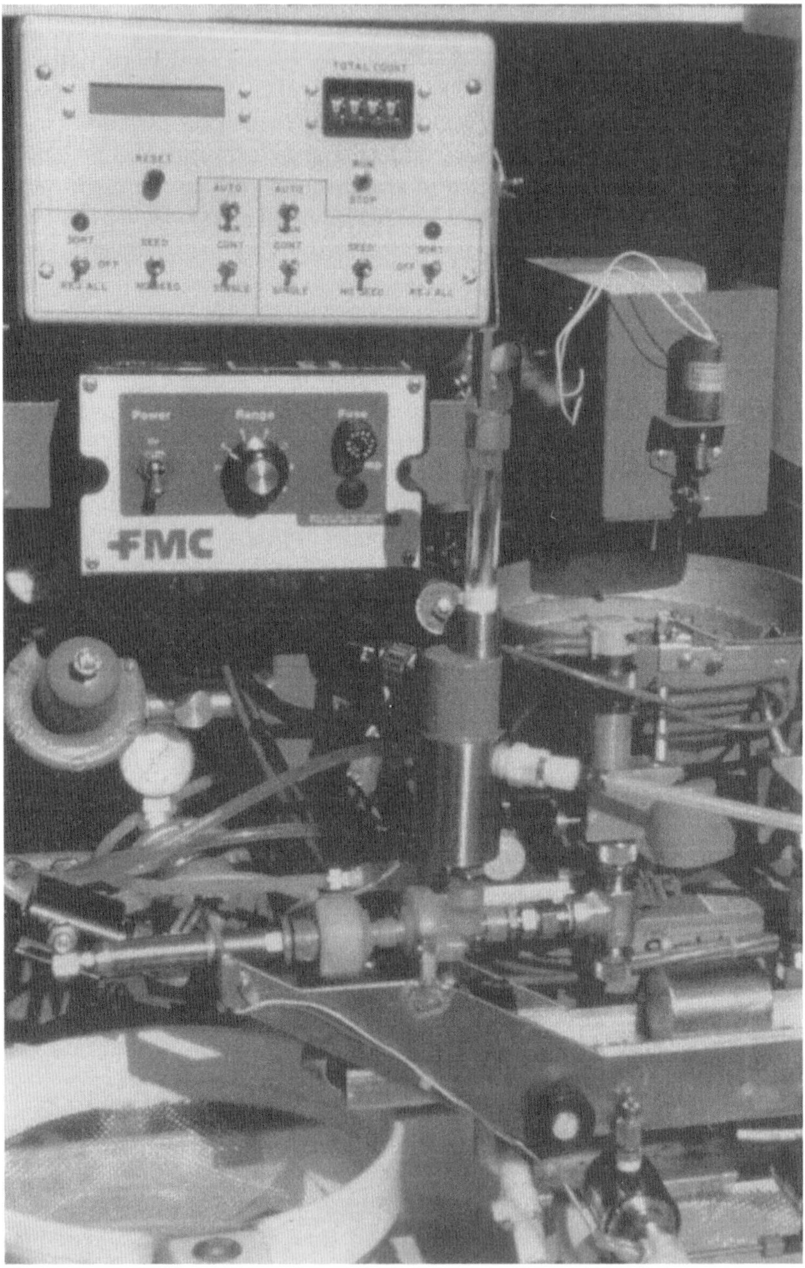

Fig. 16. Photograph of control panel of one unit which coordinates the dual ramp vibratory bowl. One nozzle and pump assembly are shown.

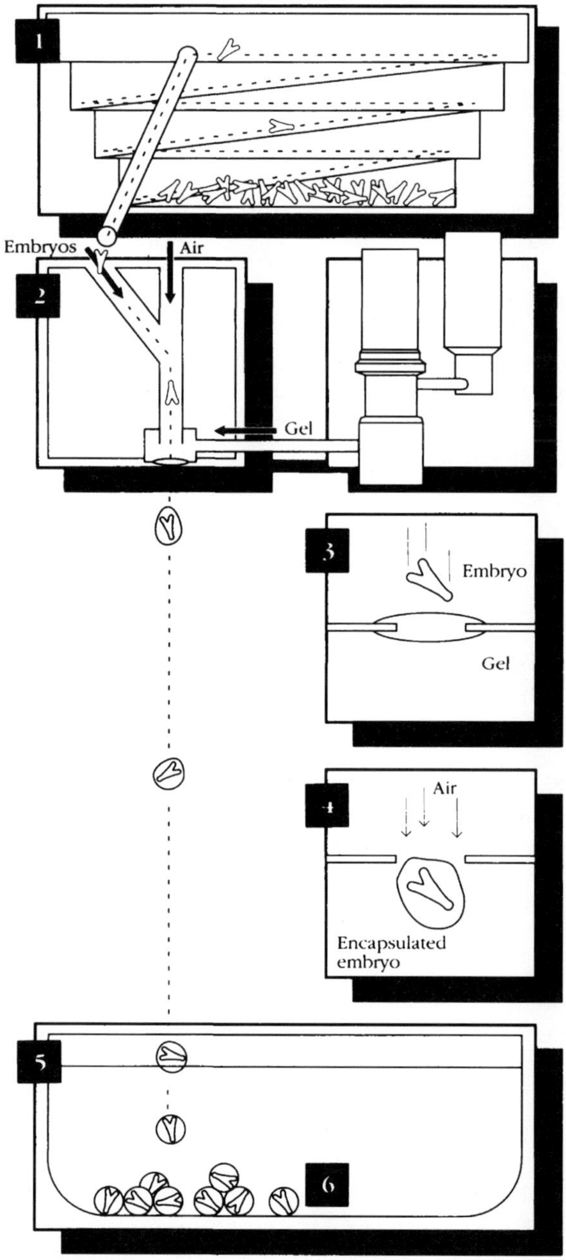

Fig. 17. (1) Embryos are metered by a vibratory bowl that has a dual spiral ramp on its inner surface. Vibration causes the embryos to climb the ramp, and at the top the width of the ramp narrows to allow only a single line of embryos to pass by a focusing, reflective *(continues)*

X. ECONOMIC CONSIDERATIONS

A. Cost Tables

Costs of alfalfa synthetic seeds have been estimated as low as $0.00026, based on development of efficient automation and bioreactor technologies (Redenbaugh *et al.*, 1987b). Recently, we calculated precise costs for producing synthetic seeds based on time and materials used in our research laboratory. This cost was $0.068, which is two orders of magnitude higher than the first (Table VIII). This large range is not unexpected since the latter figure is based on a research unit's production and the former considered economies of scale and significant mechanization. The higher cost should be considered a starting point for consideration of applicability of synthetic seed technology and the lower cost as a goal for cost reduction.

Also presented in Table VIII are the costs of direct field sowing synthetic seeds to obtain a stand establishment of 14,000 alfalfa plants. For this calculation, we assumed a very low field conversion frequency of 12%, which means that a very large number of somatic embryos (116,667) are required to produce the 14,000 plants. Again, although these figures are real numbers, they should be considered as research-level calculations.

---

infrared sensor. As long as the embryos are close together, an air jet blows them back into the bowl. Only when a single embryo passes the sensor does the air jet shut off, allowing the embryo to enter the delivery leading to the nozzle. Once the single embryo passes, the air jet is turned on again. The encapsulation cycle, which lasts approximately 1 sec, begins at the point that the embryo passes the sensor. (2) The nozzle is the heart of the encapsulation process. Embryos pass through the center tube of a coaxial tube nozzle. (3) At the beginning of a cycle, a meniscus of sodium alginate gel is stretched across the bottom end of the center tube. The amount of gel is precisely measured by a positive displacement, adjustable stroke, air pressure–driven pump. (4) Just as the embryo falls onto the alginate meniscus, an air jet puffs into the upper end of the center tube. The puff of air inflates the meniscus slightly. Gel is immediately pumped out of the outer tube with a speed sufficient to cause the drop of gel (with the embryo captured inside) to fall from the nozzle. (5) The drop of gel containing an embryo falls into a calcium chloride bath for complexing. A light source and digital camera positioned just below the nozzle detects the drop as it falls and signals for a new cycle to begin. The camera also inspects the drop. If no embryo is detected, the drop is deflected out of the path to the complexing bath and is discarded. Drops in which embryos are detected are allowed to pass. (6) The newly created synthetic seeds are collected in a basket. The computer monitors the fixing time and tells the operator when to collect the seeds for bagging. (The drawings were created by Lee Phillips, Phillips Design, Sacramento, Calif., and were used in Plant Genetics, Inc. 1989 Annual Report.)

TABLE VIII

Production Costs for Alfalfa Synthetic Seeds[a]

Planting	Efficiency	No. SE required		Cost		Person hours
Total plants	14,000					
Field conversion	12%	116,667		$0.00		0.00
Mechanical planting	50,000 capsules/hour	116,667		$0.00		2.33
Coating—Tullanox	100,000 capsules/hour	116,667	$0.000003 per capsule	$0.35		1.17
Mechanical encapsulation	7200/hr (2/sec)	116,667	$0.0004 per capsule	$50.93		16.20
Embryogenesis		**No. plates**	**No. liters medium**	**Cost**	**Labor (plates/hr)**	**Person hours**
Total embryos	116,667					
Pregermination	30 SE/plate	3,888.89	97.22	$374.52	50.00	77.78
Maturation	30 SE/plate	3,888.89	97.22	$349.83	10.40	373.93
Regeneration	40 SE/plate	2,916.67	72.92	$301.46	30.00	97.22
Induction	5 regen/induction	583.33	14.58	$51.40	15.00	38.89
Maintenance (citrate)	1.5 induction/cit	388.89	19.44	$77.80	20.00	19.44
Maintenance (SHNAAK)[b]	3 cit/SHNAAK	129.63	6.48	$22.84	20.00	6.48
Initiation	0.1 SHNAAK/init	1,296.30	64.81	$228.38	20.00	64.81
Total media and media preparation time			178.24		3.54	50.35
Misc lab supplies						
Filtration	1 unit/3 liters media	5.15/unit		$305.98		
Alcohol	1 unit/10,000 SE	5.62/gal		$65.57		
Sweater boxes	1 unit/42 plates	2.99/box		$761.34		
Parafilm	1 unit/1000 plates	7.11/box		$17.05		
Pipets, tubes, etc.	1 unit/5000 SE	10.00/5000 SE		$233.33		
Total labor		**Cost/Year**	**Years**	**Cost**		
Technician		$14,000	0.360	$5,038.76		
Total cost				$7,879.53		
Total Cost/SE				$0.068		
Total cost/field plantlet				$0.563		

[a] Assumptions: Based on RA3 protocol; encapsulation; planting using Stanhay planter; no embryos loss during pregerm, encap, planting.
[b] SHNAAK = SH medium (Schenk *et al.*, 1972), 25 μM NAA, and 10 μM kinetin (Walker *et al.*, 1981).

B. Patents

A number of patents have been issued on synthetic seed technology, for both desiccated and hydrated systems. These patents were recently reviewed (Redenbaugh *et al.*, 1990b). Considerable interest in synthetic seed technology has developed in Japan, as evidenced by a large number of applications pending. One example is synthetic seeds of carrot using alginate and carrageenan with conversion on vermiculite (Hama, 1986). Another example is coating citrus somatic embryos with a polymer such as polyvinyl alcohol, carboxymethylcellulose, gelatin, or sodium alginate and dried to 5–15% water content (Japanese Patent Appl. #JP62275604).

XI. CONCLUSIONS

Synthetic seed technology has developed considerably over the past 10 years. Improvements in somatic embryogeny have occurred such that plant conversion is possible at high frequencies *in vitro* for many species and in soil for a few. Desiccation of somatic embryos has been successful for ten crops and has the potential for providing the necessary maturation process for completing somatic embryogeny. Despite several laboratories' developing coating technology for synthetic seeds, only two have been widely tested—sodium alginate and polyoxyethylene. Additional emphasis is needed for coatings, particularly now that somatic embryo quality has been markedly improved. As the conversion rate and frequency of somatic embryos approach that of seeds, commercialization will become a reality. However, at this time, synthetic seed technology is at a research stage.

ACKNOWLEDGMENTS

We thank our co-workers at Plant Genetics, Inc. for their many contributions to synthetic seed technology.

REFERENCES

Anandarajah, K., Senaratna, T., and McKersie, B. (1989). Somatic embryogenesis in alfalfa and its use in production of dry artificial seeds. *Plant Physiol.* **89,** 8 (*suppl.*).

Anonymous. (1988a). Kirin jointly develops synthetic seed with Plant Genetics. *Kyodo News Summary* Aug. 4.

Anonymous. (1988b). Test-tube forests now a very real prospect. *Australian Timberman* Nov., p. 8

Anonymous. (1989a). Large scale, synchronous somatic embryogenesis in alfalfa. *Agricel Reports* Dec., p. 43.

Anonymous. (1989b). What are our members doing. *SAF Physiology Work. Group* August (1–2).

Baker, C. (1985). Synchronization and fluid sowing of carrot, *Daucus carota,* somatic embryos. University of Florida Masters Thesis.

Bapat, V., Mhatre, M., and Rao, P. (1987). Propagation of *Morus indica L.* (mulberry) by encapsulated shoot buds. *Plant Cell Rep.* **6,** 393–395.

Bapat, V., and Rao, P. (1988). Sandalwood plantlets from 'synthetic seeds'. *Plant Cell Rep.* **7,** 434–436.

Bapat, V., and Rao, P. (1989). *In vitro* strategies for sandalwood propagation. In "Applications of Biotechnology in Forestry and Horticulture" (V. Dhawan ed.), pp. 143–156. Plenum Press, New York.

Beunink, J., Baumgartl, H., Zimelka, W., and Rehm, H.-J. (1989). Determination of oxygen gradients in single Ca-alginate beads by means of oxygen-microelectrodes. *Experientia* **45,** 1041–1047.

Brown, D., Gibney, K., Tsekos, T., and Joy, K. (1989). Physiological markers of somatic embryo induction and development in alfalfa. *Plant Physiol.* **89,** 135 (suppl.).

Buchheim, J., Colburn, S., and Ranch, J. (1989). Maturation of soybean somatic embryos and the transition to plantlet growth. *Plant Physiol.* **89,** 768–775.

Cantliffe, D., Liu, J., and Schultheis, J. (1988). Development of artificial seeds of sweet potato for clonal propagation through somatic embryogenesis. *In* "Methane from Biomass, A Systems Approach" (W. Smith, and J. Frank, ed.), pp. 183–195. Elsevier Applied Sciences, New York.

Carman, J. (1988a). Improved somatic embryogenesis in wheat by partial simulation of the in-ovulo oxygen, growth-regulator and desiccation environments. *Planta* **175,** 417–424.

Carman, J. (1988b). Somatic embryogenesis in wheat: factors affecting competence, induction and differentiation. *In Vitro Cell. Dev. Biol.* **24,** 71A.

Carman, J. (1989). The *in ovulo* environment and its relevance to cloning wheat via somatic embryogenesis. *In Vitro Cell. Dev. Biol.* **25,** 1155–1162.

Chen, T., Thompson, B., and Gerson, D. (1987). *In vitro* production of alfalfa somatic embryos in fermentation systems. *J. Ferment. Technol.* **65,** 353–357.

Courtois, B. (1988). *In vitro* tissue culture, a tool for rice improvement. *L'Agron. Trop.* **43,** 307–316.

Daikh, H., and Demarly, Y. (1987). Resultats preliminaires sur l'obtention d'embryons somatiques et la realisation de semences artificielles de palmier dattier (*Phoenix dactylifera* L.). *Fruits* **42,** 593–596.

Datta, S., and Potrykus, I. (1989). Artificial seeds in barley: encapsulation of microspore-derived embryos. *Theor. Appl. Genet.* **77,** 820–824.

Drew, R. (1979). The development of carrot (*Daucus carota* L.) embryoids (derived from cell suspension culture) into plantlets on a sugar-free basal medium. *Hort. Res.* **19,** 79–84.

Durzan, D. (1985). Tissue culture and improvement of woody perennials: an overview. *In* "Tissue Culture in Forestry and Agriculture" (R. Henke, K. Hughes, M. Constantin, and A. Hollaender, ed.), pp. 233–256. Plenum Press, New York.

Fischer, W., Bergfeld, R., Plachy, C., Schafer, R., and Schopfer, P. (1988). Accumulation of storage materials, precocious germination and development of desiccation tolerance during seed maturation in mustard (*Sinapis alba* L.). *Bot. Acta* **101,** 344–354.

Fujii, J., Redenbaugh, K., and Walker, K. (1986). *Current status and future of artificial seed. Proc. Agriculture and Life Sciences in China.* Third Annual Ideal Colloquium. Inst. Int. Develop. and Education in Agriculture and Life Sciences. 95–111. Beltsville, Maryland.

Fujii, J., Slade, D., and Redenbaugh, K. (1988). Improving plantlet growth and vigor from alfalfa artificial seed. *In Vitro Cell. Dev. Biol.* **24,** 70A.

Fujii, J., Slade, D., Redenbaugh, K., and Walker, K. (1987a). Artificial seeds for plant propagation. *Trends in Biotech.* **5,** 335–339.

Fujii, J., Slade, D., and Redenbaugh, K. (1987b). Current status and future prospects of artificial seed. Proceedings 1987 Short Course for Seedsmen, April 13–15, 1987, Seed Tech. Lab., Miss. State Univ. pp. 13–24.

Fujii, J., Slade, D., and Redenbaugh, K. (1989). Maturation and greenhouse planting of alfalfa artificial seed. *In Vitro Cell. Dev. Biol.* **25,** 1179–1182.

Garrett, R., Mehlschau, J., and Smith, N. (1989). Gel encapsulation of tomato seeds. Amer. Soc. Agric. Engin. Paper #891573. New Orleans.

Garrett, R., Mehlschau, N. Smith, and K. Redenbaugh. (1991). Gel-encapsulation of tomato seeds. *Trans. ASAE.* In press.

Gautz, L., Upadhyaya, S., and Garrett, R. (1987a). Hydro-pneumatic singulation of gel-encapsulated propagules. Amer. Soc. Ag. Eng. Paper 87-1016. Baltimore, Maryland.

Gautz, L., Upadhyaya, S., and Garrett, R. (1987b) Vibratory separation of gel encapsulated plant propagules. *Trans. ASAE* **30,** 652–656.

Gray, D. (1987). Quiescence in monocotyledonous and dicotyledonous somatic embryos induced by dehydration. *HortScience* **22,** 810–814.

Gray, D. (1988). Effects of dehydration on quiescence and germination of grape somatic embryos. *In Vitro Cell. Dev. Biol.* **24,** 70A.

Gray, D. (1989). Effects of dehydration and exogenous growth regulators on dormancy, quiescence and germination of grape somatic embryos. *In Vitro Cell. Dev. Biol.* **25,** 1173–1178.

Gray, D., and Conger, B. (1985). Quiescence in somatic embryos of orchardgrass (*Dactylis glomerata*) induced by desiccation. *Amer. J. Bot.* **72,** 816–817.

Gray, D., Conger, B., and Songstad, D. (1987). Desiccated quiescent somatic embryos of ordhardgrass for use as synthetic seeds. *In Vitro Cell. Dev. Biol.* **23,** 29–33.

Gupta, P. (1988). Advances in biotechnology of conifers. *Curr. Sci.* **57,** 629–637.

Gupta, P., and Durzan, D. (1987). Biotechnology of somatic polyembryogenesis and plantlet regeneration in loblolly pine. *Biotechnology* **5,** 147–151.

Hama, I. (1986). Artificial seeds. Japanese Patent Application No. 40708/1986, Feb. 27, pending (in Japanese).

Hammatt, N., and Davey, M. (1987). Somatic embryogenesis and plant regeneration from cultured zygotic embryos of soybean (*Glycine max* L. Merr.). *J. Plant Physiol.* **128,** 219–226.

Hayashi, M., Ishizaki, K., Fukuoka, Y., Mashiko, T., and Sanada, M. (1988a). The morphological and developmental traits of the lettuce adventitious shoots as the encapsulated units of the synthetic seeds. *74th Annual Conf. Japan. Soc. Breed.* Oct, 38–39.

Hayashi, M., Sanada, M., Ishizaki, K., and Fukuoka, Y. (1988b). The development of the lettuce adventitious shoot multiplication system for the encapsulated units of the synthetic seeds. *74th Annual Conf. Japan. Soc. Breed.* Oct, 36–37.

Heacox, L. (1989). Seeds of change. *Ag Consultant* January, 7–8.

Jain, S. (1987). Induction and encapsulation of somatic embryoids in carrot. *Plant Phys.* **83,** 914 (suppl.).

Jain, S., Newton, R., and Tuleen, N. (1988). Tissue culture and gene transfer in barley. *Current Science* **57,** 59–70.

Janick, J. (1988). Production of synthetic seed via desiccation and encapsulation. *In Vitro Cell. Dev. Biol.* **24,** 70A.

Janick, J., Kitto, S., and Kim, Y.-H. (1989). Production of synthetic seed by desiccation and encapsulation. *In Vitro Cell. Dev. Biol.* **25,** 1167–1172.

Kaimori, N. (1988). A basic study of the development of somatic seeds. *Bull. Inst. Agr. Res. Tohoku Univ.* **39,** 21–27.

Kamada, H., Kiyosue, T., and Harada, H. (1988). New methods for somatic embryo induction and their use of synthetic seed production. *In Vitro Cell. Dev. Biol.* **24,** 71A.

Kamada, H., Kobayashi, K., Kiyosue, T., and Harada, H. (1989). Stress induced somatic embryogenesis in carrot and its application to synthetic seed production. *In Vitro Cell. Dev. Biol.* **25,** 1163–1166.

Kim, Y.-H., and Janick, J. (1987). Production of synthetic seeds of celery. *HortScience* **22,** 89.

Kim, Y.-H., and Janick, J. (1989). ABA and polyox-encapsulation or high humidity increases survival of desiccated somatic embryos of celery. *HortScience* **24,** 674–676.

King, A., Mulligan, B., and Lowe, K. (1989). Perfluorochemicals and cell culture. *Biotechnology* **7,** 1037–1042.

Kitto, S., and Janick, J. (1980). Water soluble resins as artificial seed coats. *HortScience* **15,** 439.

Kitto, S., and Janick, J. (1981). Testing artificial seed coats for asexual embryos. *HortScience* **16,** 452.

Kitto, S., and Janick, J. (1982). Polyox as an artificial seed coat for asexual embryos. *HortScience* **17,** 448.

Kitto, S., and Janick, J. (1983). ABA and chilling increase survival of asexual carrot embryos encapsulated with artificial seed coats. *HortScience* **18,** 618.

Kitto, S., and Janick, J. (1985a). A citrus embryo assay to screen water-soluble resins as synthetic seed coats. *HortScience* **20,** 98–100.

Kitto, S., and Janick, J. (1985b). Hardening treatments increase survival of synthetically-coated asexual embryos of carrot. *J. Amer. Soc. Hort. Sci.* **110,** 283.

Lawrence, R. (1981). *In vitro* cloning systems. *Environ. and Exp. Bot.* **21,** 289–300.

Li, X. (1987). Production of artificial seeds of forage carrot. *Cour. Agr. Fr.* **2,** 26–28.

Li, X., Zhu, C., Huang, M., Niu, X., Chen, B., Deng, M., and Liu, F. (1989). The production, storage and germination of high-nutrient carrot synthetic seed and mechanical screening of somatic embryos. *In* "Intern. Biotech. Symp." Peking Univ. Press, Beijing.

Lutz, J., Wong, J., Rowe, J., Tricoli, D., and Jr, R. L. (1985). Somatic embryogenesis for mass cloning of crop plants. *In* "Tissue Culture in Forestry and Agriculture" (R. Henke, K. Hughes, M. Constantin, and A. Hollaender, ed.), pp. 105. Plenum Press, New York.

McDonald, K., and Jackman, A. (1989). Bioreactor studies of growth and nutrient utilization in alfalfa suspension cultures. *Plant Cell Reports* **8,** 455–458.

McKersie, B., Bowley, S., Senaratna, T., Brown, D., and Bewley, J. (1988). Application of artificial seed technology in the production of hybrid alfalfa (*Medicago sativa* L.). *In Vitro Cell. Dev. Biol.* **24,** 71A.

McKersie, B., Senaratna, T., Bowley, S., Brown, D., Krochko, J., and Bewley, J. (1989). Application of artificial seed technology in the production of hybrid alfalfa (*Medicago sativa* L.). *In Vitro Cell. Dev. Biol.* **25,** 1183–1188.

Murashige, T. (1977). Plant cell and organ cultures as horticultural practices. *Acta Hort.* **78,** 17.

Murashige, T., and Skoog, F. (1962). A revised medium for rapid growth and bioassays with tobacco tissue cultures. *Physiol. Plant.* **15,** 473–497.

Nouaille, C., and Petiard, V. (1988). Semences artificielles: reves et realites. *Biofutur* April, 33–38.

Obendorf, R., and Slawinska, J. (1986). Somatic embryogenesis from cotyledonary tissues of soybean and maturation to desiccation tolerant state. *In Vitro Cell. Dev. Biol.* **22,** 53.

Obendorf, R., and Slawinska, J. (1988). Maturation of soybean somatic embryos to a desiccation tolerant state. *In Vitro Cell. Dev. Biol.* **24,** 71A.

Onishi, N., Kawano, T., Cooper-Smith, S., and Stuart, D. (1988a). Development of celery synthetic seed: improvement of the efficiency of somatic embryogenesis. *74th Annual Conf. Japan. Soc. Breed.* Oct, 46–47.

Onishi, N., Kawano, T., Mashiko, T., Cooper-Smith, S., and Stuart, D. (1988b). Development of celery synthetic seed: increase in germination rate of somatic embryos. *74th Annual Conf. Japan. Soc. Breed.* Oct, 44–45.

Park, H.-B., Kim, Y.-H., Whipkey, A., and Janick, J. (1988). Proline and ABA increase survival of desiccated somatic embryos of celery. *HortScience* **23,** 754.

Parrott, W., Dryden, G., Vogt, S., Hildebrand, D., Collins, G., and Williams, E. (1988). Optimization of somatic embryogenesis and embryo germination in soybean. *In Vitro Cell. Dev. Biol.* **24,** 817–820.

Redenbaugh, K. (1990a). Application of artificial seed to tropical crops. *HortScience* **25,** 251–255.

Redenbaugh, K. (1990b). Application of micropropagation to agronomic crops. *In* "Micropropagation" (P. Debergh and R. Zimmerman, eds.). In press. Kluwer Academic Publishers, New York.

Redenbaugh, K., Fujii, J., and Slade, D. (1988a). Desiccated Analogs of Botanic Seed. U.S. Patent 4,777,762.

Redenbaugh, K., Fujii, J., and Slade, D. (1988b). Encapsulated plant embryos. *In* "Advances in Biotechnological Processes" (A. Mizrahi, ed.), pp. 225–248. Alan R. Liss, New York.

Redenbaugh, K., Fujii, J., Slade, D., Viss, P., and Kossler, M. (1990a). Artificial seeds—encapsulated somatic embryos. *In* "Biotechnology in Agriculture and Forestry" (Y. P. S. Bajaj, ed.). In press. New York: Spring-Verlag.

Redenbaugh, K., Nichol, J., Kossler, M. E., and Paasch, B. (1984). Encapsulation of somatic embryos for artificial seed production. *In Vitro* **20(3),** 256–257.

Redenbaugh, K., Paasch, B. D., Nichol, J. W., Kossler, M. E., Viss, P. R., and Walker, K. A. (1986). Somatic seeds—encapsulation of asexual plant embryos. *Biotechnology* **4(9),** 797–801.

Redenbaugh, K., and Ruzin, S. (1989). Artificial seed production and forestry. *In* "Applications of Biotechnology in Forestry and Horticulture" (V. Dhawan, ed.), pp. 57–71. Plenum Press, New York.

Redenbaugh, K., Slade, D., Viss, P., and Fujii, J. (1987a). Encapsulation of somatic embryos in synthetic seed coats. *HortScience* **22,** 803–809.

Redenbaugh, K., Viss, P., Slade, D., and Fujii, J. (1987b). Scale-up: artificial seeds. *In* "Plant Tissue and Cell Culture" (C. Green, D. Somers, W. Hackett, and D. Biesboer, eds.), pp. 473–493. Alan R. Liss, Inc., New York.

Redenbaugh, K., and Walker, K. (1990b). Role of artificial seeds in alfalfa breeding. *In* "Plant Tissue Culture: Applications and Limitations" (S. Bhojwani, ed.), In press. Elsevier Sci. Publishers, Amsterdam.

Roberts, D., Flinn, B., and Sutton, B. (1989a). Desiccation promotes and synchronizes the germination of white spruce somatic embryos. *Plant Physiol.* **89,** 8 (suppl.).

Roberts, D., Flinn, B., Webb, D., Webster, F., and Sutton, B. (1989b). The role of abscisic acid and indole-butyric acid in the differentiation and quality of somatic embryos of white spruce. *Plant Physiol.* **89,** 9 (suppl.).

Roberts, O. (1990). A surprise in the test-tube. *Green Gene Gazette* Jan. 22, 8–9.

Ryan, J., and Juen, D. (1988). Crosslinked silicone coatings for botanical seeds. U.S. Patent #4,753,035.

Schenk, R., and Hildebrandt, A. (1972). Medium and techniques for induction and growth of monocotyledonous and dicotyledonous plant cell cultures. *Can. J. Bot.* **50,** 199–204.

Schultheis, J., and Cantliffe, D. (1988). Plant formation of *Ipomoea batatas* Poir. from somatic embryos in gel carriers with additives. *HortScience* **23,** 812.

Schultheis, J., Chee, R., and Cantliffe, D. (1986). Effect of growth regulators and gel carriers on growth and development of sweet potato (*Ipomoea batatas*) somatic embryos. *HortScience* **21,** 762.

Senaratna, T., McKersie, B., and Bowley, S. (1989a). Desiccation tolerance of alfalfa (*Medicago sativa* L.) somatic embryos. Influence of abscisic acid, stress pretreatments and drying rates. *Plant Sci.* **65,** 253–259.

Senaratna, T., McKersie, B., and Bowley, S. (1990). Artificial seeds of alfalfa (*Medicago sativa* L.). Induction of desiccation tolerance in somatic embryos. *In Vitro Cell. Dev. Biol.* **26,** 85–90.

Senaratna, T., McKersie, B., Bowley, S., Bewley, J., and Brown, D. (1987). Process to induce desiccation tolerance in somatic embryos. European Patent Appl. #88306589.8. July 20 filing.

Senaratna, T., McKersie, B., and Ecclestone, S. (1989b). Germination of desiccated somatic embryos of alfalfa (*Medicago sativa* L.). *Plant Physiol.* **89,** 135 (suppl.).

Shoemaker, R., and Hammond, E. (1988). Fatty acid composition of soybean (*Glycine max* (L.) Merr.) somatic embryos. *In Vitro Cell. Dev. Biol.* **24,** 829–832.

Slade, D., J. Fujii, and K. Redenbaugh. (1990). Artificial seeds: a method for the encapsulation of somatic embryos. *J. Tissue Cult. Methods* **12,** 179–183.

Slade, D., Nichol, J., Viss, P., and Stuart, D. (1990). Effect of organic acid pretreatment on the efficiency of development of whole plants (conversion) from callus cultures of alfalfa, *Medicago sativa* L. (in press).

Slawinska, J., and Obendorf, R. Soybean somatic embryo composition, respiration and water relations during maturation to a desiccation tolerant state. 1987 Annual Agronomy Meeting, Atlanta, GA (Nov. 29–Dec. 4), p. 153

Stuart, D., and Nichol, J. (1990). Effect of organic acid pretreatment on the development of somatic embryos of alfalfa, *Medicago sativa* L. (in press).

Stuart, D., and Strickland, S. (1984a). Somatic embryogenesis from cell cultures of *Medicago sativa* L. I. The role of amino acid additions to the regeneration medium. *Plant Sci. Let.* **34,** 165–174.

Stuart, D., and Strickland, S. (1984b). Somatic embryogenesis from cell cultures of *Medicago sativa* L. II. The interaction of amino acids with ammonium. *Plant Sci. Let.* **34,** 175–181.

Stuart, D., Strickland, S., and Walker, K. (1987). Bioreactor production of alfalfa somatic embryos. *HortScience* **22,** 800–803.

Styer, D. (1985). Bioreactor technology for plant propagation. *In* "Tissue Culture in For-

estry and Agriculture"(R. Henke, K. Hughes, M. Constantin, and A. Hollaender, eds.), pp. 117–130. Plenum Press, New York.
Teasdale, R., and Buxton, P. (1986). Culture of *Pinus radiata* embryos with reference to artificial seed production. *N. Z. J. For. Sci.* **16,** 387–391.
Upadhyaya, S., Gautz, L., and Garrett, R. (1987). Retrofitting vegetable planters to seed gel encapsulated propagules. *App. Eng. Agri* **3,** 211–214.
Walker, K., and Sato, S. (1981). Morphogenesis in callus tissue of *Medicago sativa:* the role of ammonium ion in somatic embryogenesis. *Plant Cell Tis. Org. Cult.* **1,** 109–121.
Wetzstein, H., Ault, J., and Merkle, S. (1989). Further characterization of somatic embryogenesis and plantlet regeneration in pecan (*Carya illinoensis*). *Plant Sci.* **64,** 193–201.
Xu, N., Coulter, K., and Bewley, J. (1989). The onset of germinability of developing alfalfa seeds. *Plant Physiol.* **89,** 168 (suppl.).
Zhong, H., and Wang, F.-X. (1989). Somatic embryogenesis in cell suspension culture of alfalfa roots and encapsulation of embryoids. *Chinese J. Bot.* **1,** 35–44.

CHAPTER **4**

Development of an Embryogenic System for Automated Propagation of Forest Trees

Pramod K. Gupta, Roger Timmis
Gerald Pullman, Mike Yancey
Mary Kreitinger, William Carlson
and Carolyn Carpenter

Technology Center
Weyerhaeuser Company
Tacoma, Washington

I. INTRODUCTION

The increasing global demand for high quality trees reflects the issues and crises created by diminishing renewable resources and increasing human populations. Unfortunately, there are still significant constraints to future tree improvement and to the productivity needed to meet this demand (Haissig *et al.*, 1987; Durzan, 1985). There is also a need to protect current resources against genetic erosion, pest damage, fire, and environmental change.

In forestry, elite traits are selected, improved, and multiplied through the use of breeding and seed orchards (Thomson *et al.*, 1987). A seed orchard is a plantation composed of vegetatively propagated trees, usually grafted, established for the production of seed of proven genetic quality. The recovery and multiplication of progeny is long term and dependent upon good seed years (Dogra, 1983). Once full seeds are developed, they must be protected from harm.

A basic strategy in tree improvement is to capture the existing genetic gains through clonal propagation. Over the past 20 years, tissue culture techniques have found wide use in commercial propagation of ornamental plants, particularly foliage houseplants. There has been some success in the use of tissue culture in the production of forest trees. However, the methods most commonly used are often labor intensive, costly, and produce low volumes. Highly mechanized culture systems able to produce large numbers of propagules efficiently must be developed in order to achieve high-frequency cloning.

Genetic gain now can be captured through somatic embryogenesis. Somatic embryogenesis offers an inexpensive, large-scale propagation system for superior genotypes. It potentially provides many advantages: large numbers (60,000 embryos/liter of medium) (Ammirato, 1986), the presence of both root and shoot meristem in the same unit, easy scale-up and transfer, opportunity for encapsulation, long-term storage, packaging, direct delivery systems, and genetic manipulation. In forestry, the production of somatic embryos throughout the year provides a complementary technology to reduce risk where seed production is limited and uncertain.

II. DEVELOPMENT OF AN EMBRYOGENIC SYSTEM

A. Culture Initiation

During seed development many conifers show multiple zygotic embryos within the seed (Singh, 1978). Multiple embryos may arise through simple polyembryony when multiple fertilizations occur and embryos form containing different genetic information or by cleavage polyembryony when the embryo cleaves to form several identical embryos. Cleavage polyembryony can be induced *in vitro,* even in species in which it is not reported in nature, such as Norway spruce and Douglas fir (Singh, 1978), and can serve as a natural rapid multiplication process.

Hakman *et al.* (1985) first reported the induction of embryogenic callus from immature Norway spruce zygotic embryos. Gupta and Durzan (1986, 1987) later described somatic embryogenesis in sugar pine, Norway spruce, loblolly pine, and Douglas fir (Gupta and Durzan, 1986, 1987). They termed this embryogenic tissue "embryo suspensor masses" due to its high degree of organization. A number of other coniferous species have been reported to produce embryogenic tissue *in vitro* (Table I). Embryonal suspensor masses of conifers have a distinctly different phenotype compared with nonembryogenic callus. Embryonal suspensor masses (ESM) are usually white, translucent, and moist or mucilagenous compared with nonembryogenic callus, which often appears opaque, friable, and may turn green in the light (Hakman *et al.*, 1985; Gupta and Durzan, 1986; Krogstrup, 1987).

There appear to be two methods for the initiation of embryogenic cultures of *Picea:* through the continuation of natural cleavage polyembryony in cultured immature embryos (Hakman *et al.*, 1985), and through the induction of callus from the hypocotyl or cotyledon, which rapidly organizes to form embryos which further multiply through cleavage polyembryony (von Arnold and Hakman, 1987; Nagmani *et al.*, 1987). All reports of initiation of Douglas fir or *Pinus* spp. embryogenic cultures are based on the use of immature embryos. Even in cultured mature seeds of sugar pine, ESM arose from the suspensor region, presumably from undeveloped proembryos by cleavage polyembryony (Gupta and Durzan, 1986).

Media requirements for the initiation process do not appear to be very specific. Our own experience at Weyerhaeuser Company shows that Douglas fir ESM can be induced on many different basal salts such as DCR (Gupta and Durzan, 1985), modified MS (Murashige and Skoog, 1962), P6 (Teasdale *et al.*, 1986), or Litvay Medium (Litvay *et al.*, 1985). The auxin 2,4-dichlorophenoxy acetic acid (2,4-D) generally has been the preferred auxin for initiation of ESM in *Picea abies* (von Arnold, 1987; Lelu *et al.*, 1987).

ESM induction has been carried out with different concentrations of hormones, such as up to 11 mg/liter 2,4-D (Hakman *et al.*, 1985; Gupta and Durzan, 1986, 1987), up to 5 mg/liter naphthelene acetic acid (von Arnold, 1987; Lelu *et al.*, 1987) and up to 5 mg/liter kinetin or benzyl aminopurine (von Arnold and Hakman, 1987; Hakman and Fowke, 1987; Gupta and Durzan, 1986). In Norway spruce ESM induction occurs

TABLE I
Somatic Embryogenesis in Conifers

Species	Explants	Response	Established in soil	Reference
Norway spruce				
Picea abies	IME	ESM, PL	No	Hakman *et al.* (1985)
			No	Hakman & von Arnold (1985)
			Yes	Becwar *et al.* (1987) [1989]
	ME	ESM, PL	Yes	von Arnold & Hakman (1986) [1988]
			Yes	Gupta & Durzan (1986)
			Yes	von Arnold (1987)
			No	Boulay *et al.* (1988)
			No	Jain *et al.* (1988)
			Yes	Verhagen & Wann (1989)
	COT	ESM, PL	No	Krogstrup (1986)
			Yes	Lelu *et al.* (1987)
	ME	ESM, PL	Yes	Gupta *et al.* (1989)
White spruce				
Picea glauca	IME	ESM, PL	No	Hakman & Fowke (1987)
			Yes	Lu & Thorpe (1987)
			Yes	Dunstan *et al.* (1988)
			Yes	Webb *et al.* (1989)
			No	Nagmani *et al.* (1987)
	ME		Yes	Trembelay (1988)
			Yes	Hakman & von Arnold (1988)
			Yes	Webster & Roberts (1989)
Picea engelmannii	IME	ESM	No	Webb *et al.* (1989)
Black spruce				
Picea mariana	IME	ESM, PL	No	Hakman & Fowke (1987)
Sitka spruce				
Picea sitchensis	IME, ME	ESM, PL	No	von Arnold & Woodward (1988)
			Yes	Krogstrup *et al.* (1988)
Larch				
Larix decidua	FG	ESM, PL	No	Nagmani & Bonga (1985)
Hybrid larch				
decidua × *leptolepsis*	IME	ESM, PL	No	Klimaszewska (1989)
Sugar pine				
Pinus lambertiana	IME, ME	ESM, PL	No	Gupta & Durzan (1986)
Loblolly pine				
Pinus taeda	IME	ESM, PL	Yes	Gupta & Durzan (1987)
			Yes	Durzan & Gupta (1988)
			Yes	Gupta & Durzan (1989)
			Yes	Gupta *et al.* (1989)

(*continued*)

TABLE I ***(continued)***

Species	Explants	Response	Established in soil	Reference
Pond pine				
Pinus serotiana	IME	ESM	No	Becwar *et al.* (1988)
White pine				
Pinus strobus	IME	ESM	No	Becwar *et al.* (1988)
			No	Finer *et al.* (1989)
Hybrid pine				
P. taeda × *P. resinosa*	IME	ESM	No	Becwar *et al.* (1988)
Radiata pine				
Pinus radiata	IME	ESM, PL	Yes	Smith *et al.* (1989)
Slash pine				
Pinus ellottii	IME	ESM	No	Jain *et al.* (1989)
Douglas fir				
Pseudotsuga menziesii	IME	ESM, PL	Yes	Durzan & Gupta (1987, 1988)
			Yes	Gupta *et al.* (1989)
	IME	ESM	No	Webb *et al.* (1989)
Redwood				
Sequoia sempervirens	IME	ESM, PL	No	Bougkard & Fave (1988)
White fir				
Albies alba	IME	ESM	No	Schuller *et al.* (1988)

COT, cotyledon; ESM, embryonal suspensor mass; FG, female gametophyte; IME, immature embryo; ME, mature embryo; PL, plantlet; SS, somatic seedling.

more frequently in media with lower levels of sucrose (1.5–2% as opposed to 3%) (Becwar *et al.*, 1987). ESM induction most often occurs in the dark but has also been reported to occur in the light (Verhagen and Wann, 1989).

ESM induction has been reported in immature embryos, mature embryo hypocotyls, and germinated cotyledons of Norway spruce (Hakman *et al.*, 1985; Krogstrup, 1987; Lelu *et al.*, 1987). Immature and mature embryos have also proven to be effective explants for white spruce and sitka spruces (Table I). In other coniferous species, success has been achieved mostly with immature explants (Table I). Mature embryos of sugar pine also produced ESM but at a very low percentage (0.1%) (Gupta and Durzan, 1986). It thus appears that among conifers the genus *Picea* is unique in the ability to initiate somatic embryos from hypocotyl or cotyledonous tissue. However, a short callus stage occurs where the tissue reorganizes to form the first somatic proembryo. Subsequent somatic proembryos develop through cleavage polyembryony. The short callus stage increases the risk of somaclonal variation. Recent reports indicate that variation exists in ESM started from cotyledons (Lelu, 1989). Although attempts have been made to initiate ESM from mature tissue, none has been successful.

B. Culture Maintenance

Once an ESM is established it must be maintained in culture without loss of morphogenetic competence. Many authors fail to point out that initiated ESM frequently discontinues growth and the culture is lost. Cultures initiated from immature *Picea glauca* embryos showed a loss of 50–75% in a few months (Webb *et al.*, 1988). A 60% loss was reported for cultures from mature embryos of *P. stichensis* (von Arnold and Woodward, 1988). More than 50% of the cultures started from immature Douglas fir embryos had discontinued growth within 6 months (D. Webb, personal communication).

Cultures are usually maintained at lower levels of hormones than required for initiation. About 2 mg/liter 2,4-D and 0.5 ppm cytokinin (Kinetin or BAP) are typical (Gupta and Durzan, 1986, 1987). Frequent subculture, every 10–12 days on solid medium and every 7 days in liquid medium, is necessary to reduce loss of embryogenic potential and avoid callus formation. Cultures are most often maintained in the dark at 22°–25°C. Conifer embryo multiplication occurs through the natural process of cleavage polyembryony (Gupta and Durzan, 1986, 1987). On solid medium, colonies often double in volume every 2 weeks, while in suspension culture, cell mass increases 2- to 4-fold every 7 days. For the latter (see Color Plate 1) cell density is critical and often determines the quality of the proembryos formed. For example, ESM densities of 5 ml packed cell volume in 40 ml fresh medium were required for maximum embryo quality in Douglas fir (Durzan and Gupta, 1987). Lower densities led to insufficient growth, while higher densities caused browning of cells.

C. Embryo Development and Maturation

For further embryo development, cleavage polyembryony must be inhibited through the removal of auxins and cytokinins and the addition of abscisic acid (ABA). Several methods are currently used to support continued development of the embryos. Proembryos may be transferred

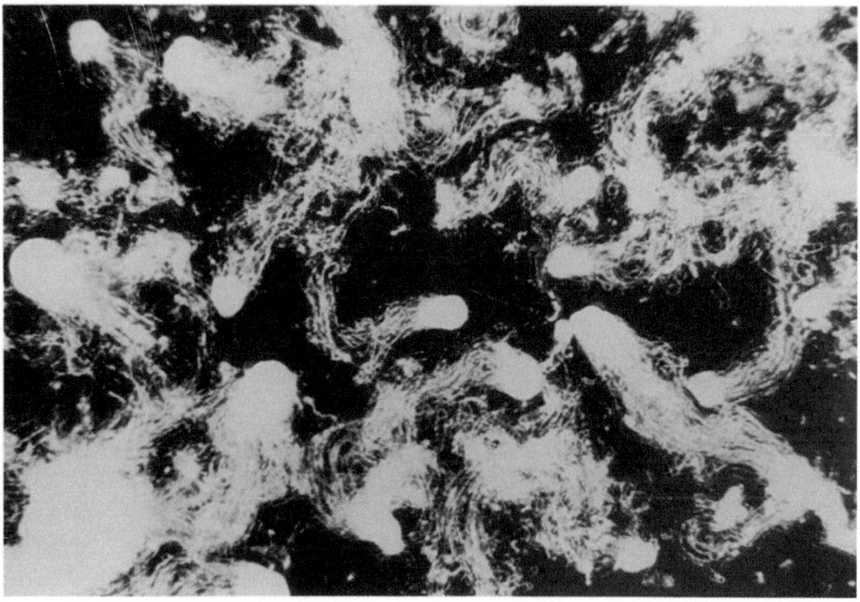

Color Plate 1 Embryonal cell suspension culture of Norway spruce (*Picea abies*).

Color Plate 2 Somatic embryos of Norway spruce (*Picea abies*) growing on semisolid medium after ABA treatment.

Color Plate 3 Germinating somatic embryos of Norway spruce (*Picea abies*) on semisolid medium.

Color Plate 4 Converted plantlets from somatic embryos of Norway spruce (*Picea abies*) established in soil.

Color Plate 5 Norway spruce (*Picea abies*) somatic seedlings in soil in a greenhouse. More than 4,000 plants from 17 genotypes have been established in soil.

Color Plate 6 Automated Mini-Plug^{TM} transplanter planting Douglas fir (*Pseudotsuga menziesii* Mirb.) seedlings from mini-plug trays to soil.

Color Plate 7 Somatic embryos of Norway spruce (*Picea abies*) encapsulated in calcium alginate gel.

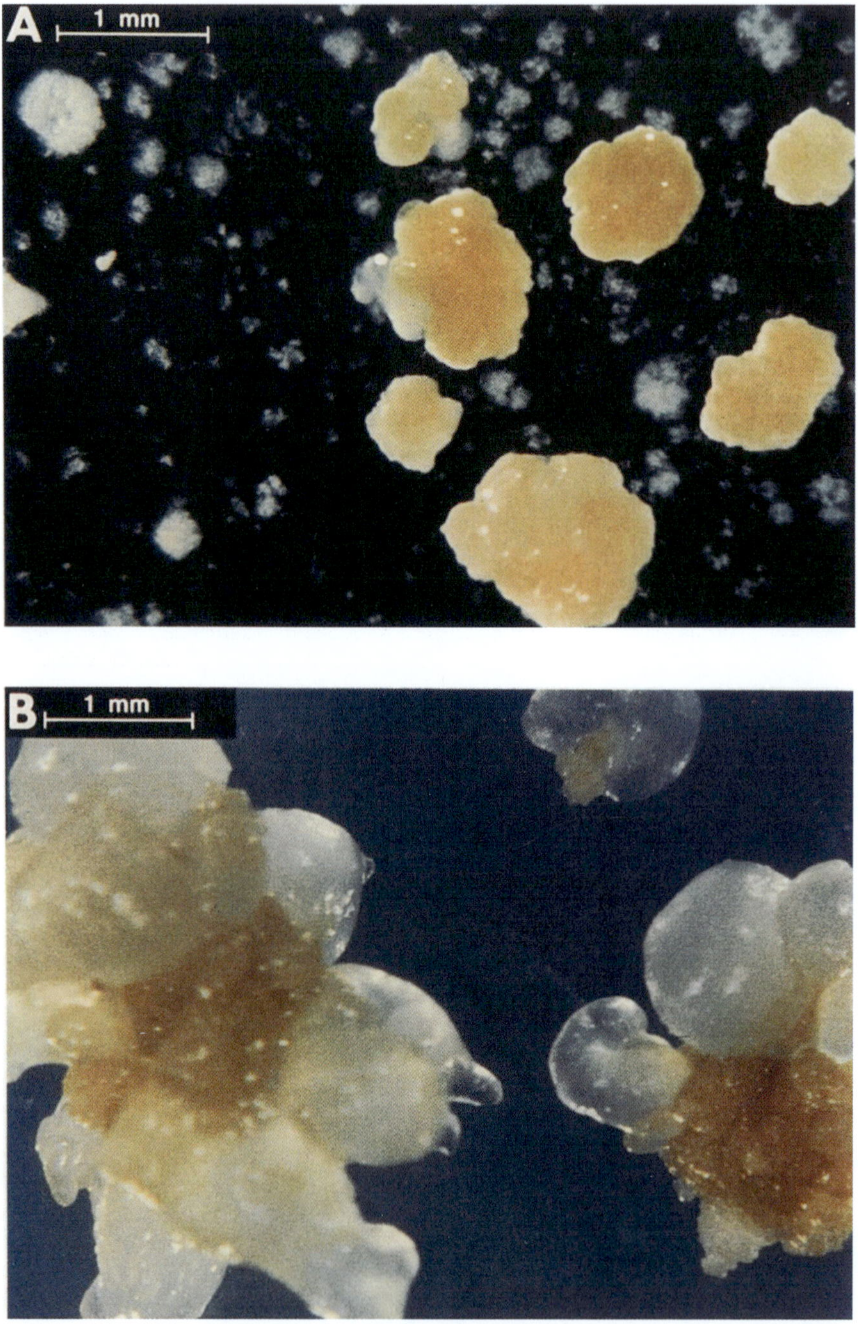

Color Plate 8 Embryo production of *Ipomoea batatas* in suspension culture. (A) Various sizes of embryogenic callus aggregates grown on medium containing 2,4-D. (B) Embryo clusters forming on media without 2,4-D.

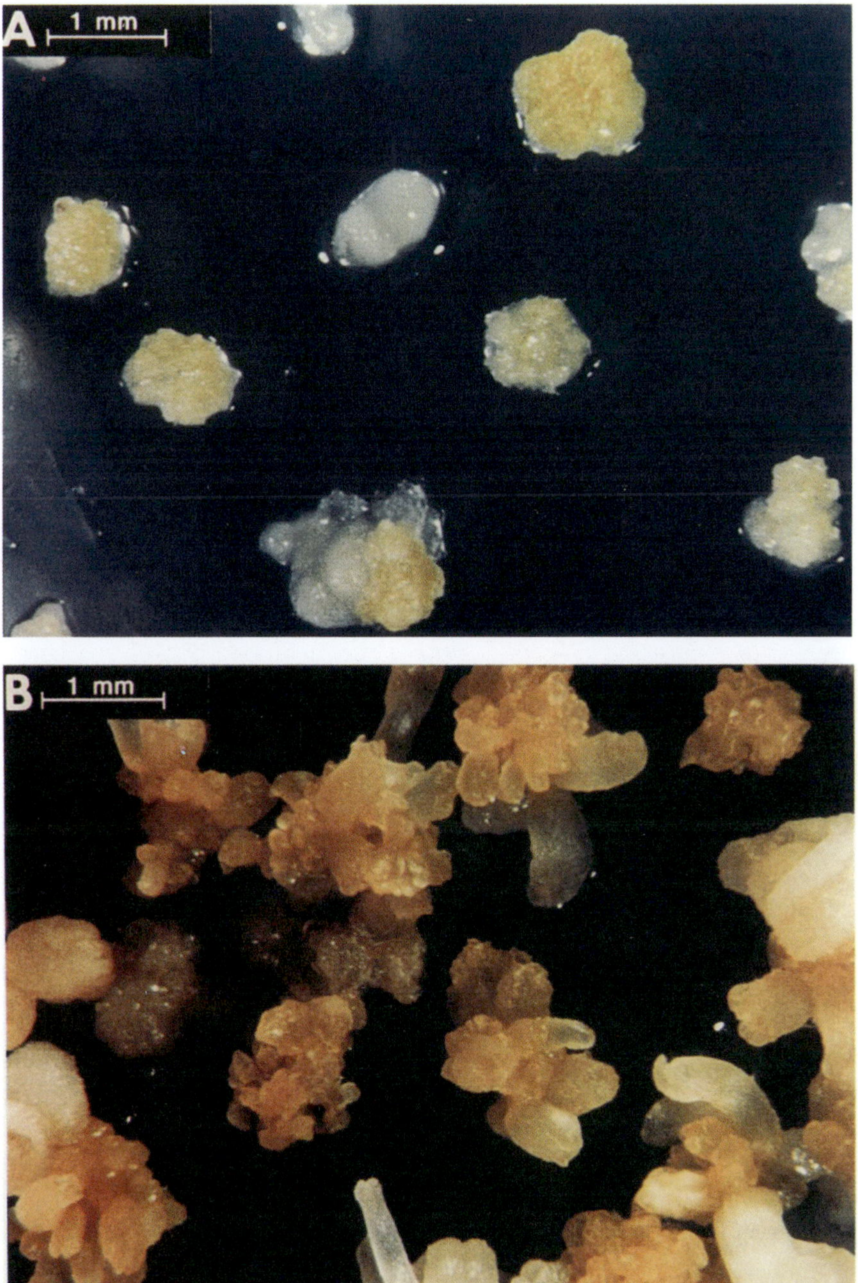

Color Plate 9 To study callus growth rates 710–1000 μm calli subunits were used. (A) Calli clusters containing both yellow embryogenic and white nonembryogenic callus. (B) 710–1000 μm calli subunits grown in media without 2,4-D for 18 days showing embryo clusters forming.

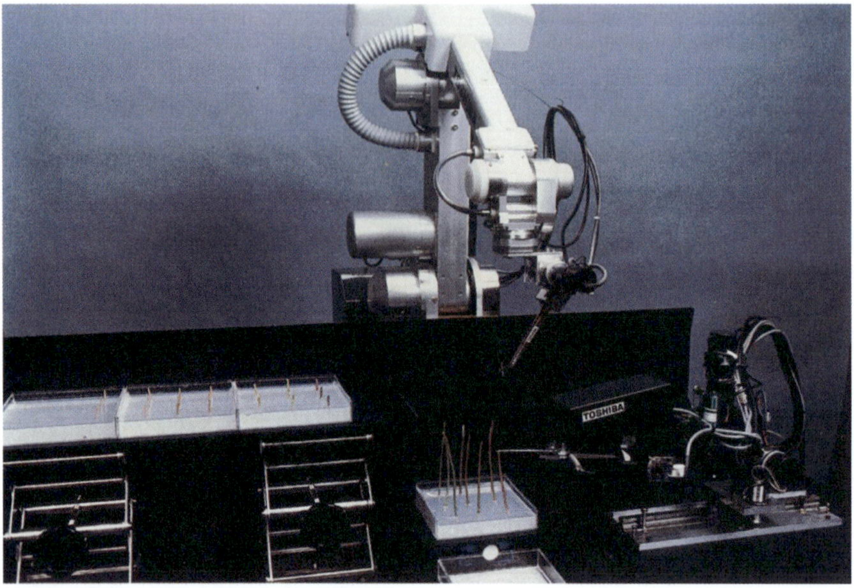

Color Plate 10 A view of the prototype robot system.

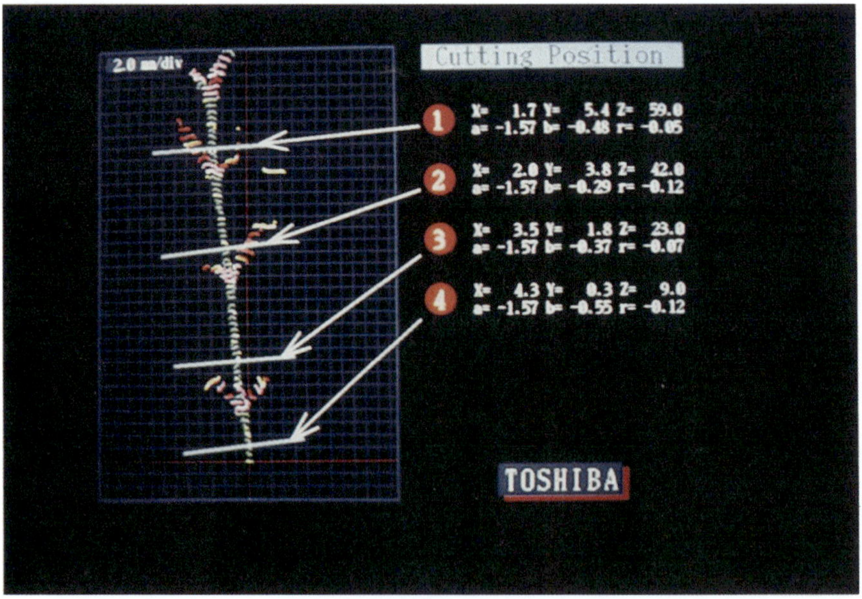

Color Plate 11 An example of node recognition and cutting position.

to a medium containing activated charcoal and then exposed to ABA (Becwar *et al.*, 1987) or directly transferred to ABA-containing medium for several subcultures (Durzan and Gupta, 1987; von Arnold and Hackman, 1988). ABA has also been used with indole-butyric acid (IBA, 0.1–1 ppm) for development of Norway spruce somatic embryos (Becwar *et al.*, 1987).

Abscisic acid plays several important roles in conifer embryo development and has been used at concentrations of 0.1–15 ppm in different coniferous species (David, 1989). Cleavage polyembryony is inhibited by ABA, allowing singulation and continued growth of individual embryos. In addition, ABA plays a role in embryo maturation and storage compound accumulation. Recently, complex carbohydrates have been combined with ABA for improved embryo development and maturation. For example, 6% sucrose and ABA were useful for loblolly and white pines (Finer *et al.*, 1989; Becwar and Feirer, 1989), and maltose, lactose, and starch for white fir embryo development (Schuller and Reuther, 1989).

Further embryo development often occurs with the removal of ABA by activated charcoal (Durzan and Gupta, 1987) or transfer to medium free of ABA (Boulay *et al.*, 1988). Embryo development has been obtained on a semisolid medium (see Color Plate 2) or on artificial supports such as filter paper, cheesecloth, or fabric saturated with liquid medium (Gupta and Durzan, 1986; Boulay *et al.*, 1986; von Arnold and Hakman, 1988).

Embryo development media often contain reduced nitrogen in the form of amino acids such as glutamine, asparagine, and arginine (Boulay *et al.*, 1988). Gelrite, a bacterial gellan gum used as a gelling agent, has been reported to produce more embryos than agar (Hakman and von Arnold, 1988). Successful cotyledonary embryo development has been reported in many coniferous species (Table I) using different protocols.

D. Genotypic Variation

A high degree of genotypic variation has been observed in all stages of embryo development. In loblolly pine one out of ten cultures produced cotyledonary embryos (M. R. Becwar, personal communication). In Douglas fir, one out of six cultures produced cotyledonary embryos. However, when superior full-sib families were used for culture initiation in Douglas fir, ESM from the same families showed similarities in initiation frequency, maintenance difficulty, proembryo size, and embryo development (Gupta *et al.*, 1989).

III. PLANTLET REGENERATION AND ESTABLISHMENT

A. Germination

Germination of somatic embryos has been reported in a number of coniferous species (Becwar *et al.*, 1989), often exceeding 35–65%. However, germination is often described as enlargement of cotyledons, elongation of the hypocotyl region, and growth of the primary root. Often there is no mention of epicotyl growth, which is critical for continued embryo survival.

Most germination media contain half strength salts, 1–2% sucrose, activated charcoal, *myo*-inositol, vitamins, and a low level of reduced nitrogen as amino acids. Norway spruce embryos germinated best after 7 days in the dark followed by continuous light. In our laboratory we have obtained 80–90% germination from selected quality embryos on semisolid medium (see Color Plate 3).

B. Conversion

Conversion is the term used to denote the process of turning somatic embryos into plants growing autotrophically in soil. With few exceptions, the papers on conifer embryogenesis cited herein report only small numbers of plants successfully converted. Conversion typically is achieved by germinating selected well-formed embryos on a sucrose germination medium until true leaves appear and transplanting them into soil. Care is taken to provide a humid atmosphere (e.g., by using plastic covers or a greenhouse fog system) and to gradually acclimate the plants to lower humidities before attempting to move them outdoors.

Only two reports describe attempts to convert large numbers of conifer embryos from a sufficient number of genotypes to provide meaningful yield data or enough material for field testing. Webster and Roberts (1989) converted 1,600 somatic embryos of white spruce making up 71 genotypes. Conversion was considerably improved (from 14 to 65%) by a mild desiccation treatment at 97% relative humidity (Roberts *et al.*, 1989).

In our laboratory we have converted some four thousand Norway spruce embryos, making up 17 genotypes, to plants (see Color Plates 4 and 5). Selected embryos are like zygotic embryos and will "hold" on ABA-containing media without germinating or deteriorating for many weeks. In most genotypes 80–90% of such embryos will germinate in a fairly synchronous manner on hormone-free medium without further treatment (Gupta *et al.*, 1989). More than 200 normal-looking "somatic seedlings" of Douglas fir have also been produced from a total of six genotypes. In this species, however, the technology for embryo development and maturation is less advanced, and the conversion efficiency per 100 embryos is still considerably lower than for Norway spruce.

The most important requirement for conversion is to start with fully developed and matured embryos. Such embryos must not only have normal external appearance but also must have internal characteristics approximating those of zygotic embryos. These include storage compounds—protein, lipid, and carbohydrate—and an essentially normal tissue structure. Based on findings in agronomic crop species (e.g., Buchheim *et al.*, 1989; Kermode *et al.*, 1989), these characteristics are likely to be associated with an ability to tolerate desiccation to <10% water content, which itself is a property of potential value for storage and delivery systems.

C. Quality Control

Quality of somatic embryos is very important for successful germination, conversion to plant, encapsulation, or automated delivery. The development of protocols for high-quality somatic embryos has been enhanced through the use of biochemical and cytological techniques. Measurements of abscisic acid, seed storage proteins, fatty acids, and carbohydrates in somatic and zygotic embryos have helped to improve somatic embryo quality. In addition, valuable information on somatic embryo anatomy has been obtained through microtechnique. With the help of such information, we have produced high-quality somatic embryos of Norway spruce, Douglas fir, and loblolly pine.

IV. CRYOPRESERVATION

Cryopreservation is very important for the large-scale commercial use of somatic embryogenesis. Clonal field testing in forestry often takes 5–10 years before superior genotypes are identified. ESM provides an excellent cell type for storage of juvenile tissue during this field testing period. Once desirable genotypes are identified, they can be retrieved from liquid nitrogen and scaled up through cleavage polyembryony. In addition, cryopreservation provides an opportunity for creation of a bank of superior genotypes which can be retrieved when needed in the commercial operation. Successful short-term storage in liquid nitrogen and plant regeneration have been demonstrated for Norway spruce, white spruce, and loblolly pine (Gupta *et al.*, 1987; Galerine and Dereuddur, 1987; Kartha *et al.*, 1988). For commercial application, a high proportion of desirable genotypes (versus cultures of a given genotype) must survive cryopreservation. No data on this aspect of cryogenic storage yields are yet available.

V. CULTURE SCALE-UP

Cell suspension cultures of 50–300 ml can easily be maintained in 250–2000 ml Erlenmeyer flasks. Cultures are incubated on rotary shakers at 80–120 RPM at 25°C in the dark. Nipple flasks have also been used to scale up embryogenic suspension cultures (Durzan and Gupta, 1987). Recently, a 10-liter bioreactor was used to scale up *Pinus radiata* cell suspensions (Smith, 1989). However, in these bioreactor experiments, only 5–6 cotyledonary embryos were reported to develop when plated on semisolid medium. Until now cotyledonary embryos have not been produced in immersed liquid systems. Usually proembryos require plating onto a semisolid medium or artificial support system after development in a liquid medium. Twenty to 30 cotyledonary embryos of Norway spruce have been reported to form from 1 ml of suspension culture when placed on cheesecloth or filter paper supports wetted with liquid medium (Boulay *et al.*, 1988). In our laboratory, 50–100 cotyledonary embryos of Douglas fir or Norway spruce develop from 1 ml of suspension culture plated on semisolid medium. Occasional batches from 2-liter flasks have produced 60–70 ml of settled cells, which when plated produced approximately 3,000 embryos, demonstrating the large-scale capabilities of somatic embryogenesis. This translates to over 50,000 cotyledonary embryos per liter of cell suspension.

VI. DELIVERY SYSTEMS

A. Containerized

It is likely that the first commercial application of embryogenesis in forestry will utilize modifications of existing container systems. Later, these may be supplanted by direct sowing of "artificial seed" into nurseries.

Such a system may use machine-assisted or automated sowing of pregerminated embryos into containers. Supplementary nutrient gel encasement, as in fluid drilling technology (Gray, 1980), modified for greenhouse use, may shorten the conversion period. Key biological factors to be considered are: (1) management of any *ex vitro* heterotrophic growth in order to minimize microbial contamination and speed the onset of photosynthesis, and (2) humidity control of the greenhouse environment to allow acclimatization of transplanted plantlets.

Containerized somatic seedlings would be machine planted in a nursery for a second season's growth. Automated transplanting systems have been developed for vegetable growers by such companies as Grower's Transplanting, Inc. (GTI) and Castle and Cooke Development Company of the United States, Williams Hi-Tech International, Ltd. of Australia, and Lannen Tehtaat Oy of Finland (Chitwood, 1989).

Weyerhaeuser Company has licensed GTI's vegetable sowing systems, modified it for use with forest tree seedlings, and given it the name Mini-Plug™ Transplanting System (see Color Plate 6). One of its unique features is that the trays are bottomless. Pneumatic spears impale the plant-bearing cells, push them out of the bottom of the tray, and guide them throughout the plug-setting process (Branch, 1987). This and other systems could be adapted readily to the products of embryogenesis.

We can also envisage a range of hybrid delivery systems utilizing aspects of both containerized and artificial seed technologies. For example, crudely encapsulated embryos may be delivered via a greenhouse container system or into environmentally modified nursery beds (e.g., with plastic row covers) at lower overall cost. To date, no reports of modified conventional delivery systems for somatic embryos have appeared in the scientific or patent literature.

A further requirement of large-scale delivery systems is the capability for short-term storage of embryos or resulting plants. This will be needed as a buffer between laboratory production, which should be continuous to minimize costs, and field outplanting, which is seasonal. Our preliminary data for Norway spruce indicate that it will be possible to meet this need most easily by accumulating embryos either in a cold room or in a highly desiccated state, or both. Synchronized germination can then be achieved after imbibition and culture under favorable conditions.

B. Artificial Seed

Artificial seeds of high conversion potential would allow interfacing of somatic embryo production with conventional containerized or bare-root nursery production. Several patents exist concerning artificial seed design, but none of the patented technology yields a high percentage of normal germination (Redenbaugh *et al.*, 1985; see also Chapter 3, this volume). Germination is limited by lack of supply of nutrients, lack of sufficient oxygen supply, and by limited tolerance to desiccation, microbial invasion, and mechanical damage. Norway spruce (see Color Plate 7) loblolly pine, and radiata pine have been encapsulated in hydrated calcium alginate gel (Durzan and Gupta, 1988; Smith, 1989). However, only 1–2% germination has been achieved with encapsulated embryos.

The rate of development of concepts and methods for producing artificial seeds is rapid, requiring research on both biology and materials science. In the future, production facilities producing somatic embryos as artificial seed may replace production seed orchards. However, breeding programs must continue to produce new genotypes to maintain diversity.

VII. SINGULATION AND SELECTION OF EMBRYOS

The labor cost required to individually select and move embryos is relatively high. Therefore, some degree of mechanization will be necessary at this stage. There has been no reported development of an integrated, automated system for singulating and selecting conifer embryos. Taking a broader look across all plant species, the only reported attempt at developing such a system is the "bioprocessor" component of the Vitromatic System of Plant Biotech Industries in Ashrat, Israel (Levin *et al.*, 1988). Individual components that might have a place in an integrated system are being developed by various researchers around the world.

Five key steps are involved in singulating and selecting conifer somatic embryos: (1) removal from the embryo development media, (2) separation from the ESM (embryo suspensor mass), (3) sorting of the embryos, (4) singulation of the embryos, and (5) selection of the embryos. The above activities must be carried out in a sterile environment since media containing sugars will be required during germination and initial growth.

The method chosen for embryo removal will depend on the type of embryo development system being utilized. If a liquid system (e.g., bioreactor) is being used, the embryos can be drawn off or piped away from the main vessel. If a solid system (e.g., agar) or semisolid system (e.g., pad) is being used, the embryos can be washed off the surface with a gentle stream of sterile water. In all cases, the ESM will still be attached to the embryos and will travel with them.

There are at least two approaches for separating embryos from the ESM. One would involve agitating the embryos and ESM in sterile water to break them apart, then washing the ESM through a sieve or screen, leaving the embryos behind. Another would involve agitating the embryos and ESM in a sterile liquid of chosen specific gravity. After separation, the ESM would float to the surface and be skimmed off while the embryos would sink.

Embryo sorting can be carried out by size fractionation using screens or sieves (Lomax *et al.*, 1988; Warren and Fowler, 1977), density fractionation using a discontinuous percoll density gradient (Lutz *et al.*, 1985; LoSchiavo, 1984), or velocity sedimentation using a centrifuge (LoSchiavo, 1984).

Singulation of the remaining embryos can be accomplished in pipe flow by proper choice of pipe diameter, surface drag characteristics, and liquid velocity. Another approach utilizes a vacuum or suction pickup to move individual embryos out of a recirculating liquid bath.

Embryo selection will also involve careful morphological inspection and selection based on the external characteristics. As each embryo moves through an inspection station, it would be optically magnified in size, evaluated by the operator, and selected or rejected. In a more advanced and automated system, image analysis (e.g., machine vision) could be used to automatically inspect and accept or reject each embryo (Grand d'Esnon *et al.*, 1989; see also Chapters 8 and 9, this volume).

VIII. FIELD TESTING AND IMPLEMENTATION

The ultimate limit to the implementation of somatic embryogenesis technology for clonal forestry probably will not be our ability to develop the technology itself but rather the time required for field testing the product. This is especially true in temperate regions. For example, a minimum of 6 years in the field is considered necessary for Douglas fir seedling family ranks to stabilize in genetic tests. In Weyerhaeuser's tree improvement program for this species, final selections are not made until trees are 10–12 years old. This leads to an implementation/genetic gain timeline like that in Fig. 1.

The base case in Fig. 1 is the seed production orchard. Let us suppose that at time zero on the *x*-axis the (superior) trees in this orchard begin to bear flowers. Only a very few are produced at first. The resulting seed provides only enough genetically improved trees 2 years later to reforest a small proportion of the land. This is the first stairstep, shown in year 3. The remainder is reforested with wild (unimproved) seed.

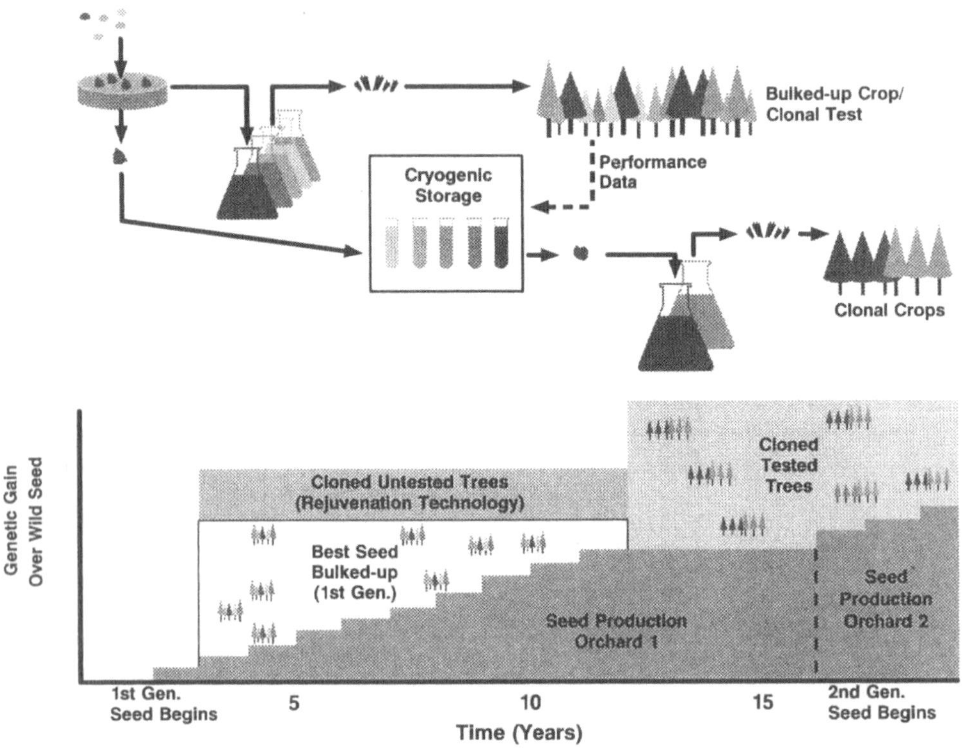

Fig. 1. Use of somatic embryogenesis in forestry. Embryos from elite family seed, which is in short supply, are multiplied in liquid culture. The cloned embryos (perhaps as artificial seed) are used: (1) to "bulk up" planting stock so that the shortfall in seed orchard production can be eliminated, allowing additive genetic variance gain to be captured on all reforested land almost immediately, and (2) to establish clonal tests. Some 5 years later, performance data from these tests are used to select embryonal suspensor masses from cryogenic storage for production of clonal plantations. These plantations will have higher value due to the capture of the nonadditive portion of genetic variance. The cycle will continue with subsequent sexual generations.

Each subsequent year, more seed is produced until the orchard reaches full production (year 12 in Fig. 1). At this point all available land can be reforested with improved first-generation seed. Several years later the second-generation orchard comes on line (established from the best field-tested progeny of the first orchard), and the stairstep of increasing genetic improvement resumes. This is current technology.

Embryogenesis technology can be applied to enhance this base case in three ways. First, we could remove the immature embryos from the first cone crop of the production orchard, initiate and multiply ESM, and produce enough somatic embryos from this seed mix to make up the shortfall. Termed "bulking up," this process would give us the full first-generation genetic gain 8 years sooner—a tremendous economic value but due only to a compression of the base case timeline. In fact, the gain itself could be increased by bulking up a select subset of the orchard families.

A second possibility would be to get material from the first genetic tests that were established from controlled crosses among the seed orchard trees. At Weyerhaeuser Company, these crosses were made among trees that were induced by special treatments to flower at a much younger age than the rest of the orchard trees. Consequently the progeny are already old enough to be evaluated for growth and form, and the best among them are truly outstanding individuals. Cloning these would probably give a genetic gain beyond that from bulked-up seeds whose individual performance has never been observed. One problem with this, however, is our current lack of a "rejuvenation" technology for initiating ESM from adult trees, and a greater risk of genetic variants should an extension of the cotyledon induction technology of Lelu *et al.* (1987) be developed and used. A second problem is that, although we can see that the individuals are superior, we do not know the extent to which the superiority is genetic versus environmental (because the individual seedlings are not replicated). Therefore, to fully exploit it would require a clonal test period exceeding 5 years.

The third way to apply embryogenesis technology to the base case tree improvement program is illustrated by the flow diagram occupying the top portion of Fig. 1. This repeats the bulking-up process described above but saves a portion of the ESM from each successful initiation of orchard seed in cryogenic storage. The performance of each clone is tested on several sites over several years until the best clones can be selected with confidence. ESMs for those clones are then recovered from cryostorage and multiplied to the desired extent. This approach has the advantage that the required technology is close to being available and genetic gain is maximized. However, a several-year wait for performance information is unavoidable. Only after the clonal selections have been made can embryogenesis technology be fully utilized.

IX. GENETIC DIVERSITY

To maintain stability of the forest in the face of risks from pests and climate extremes a multiplicity of clones will surely be used (Huhn, 1986, and papers cited therein). Whether or not they should be deployed as single clones in a mosaic of small blocks, as Aracruz does with eucalyptus in Brazil (Zobel et al., 1983), or as a mixture of clones within a plantation, remains an important research and management question whose answer may vary depending on species and geography. Developing a responsible strategy for the deployment and management of clones is essential. At Weyerhaeuser Company, this effort is advancing in parallel with development of the somatic embryogenesis technology itself.

REFERENCES

Ammirato, P. V. (1986). Carrot. *In* "Handbook of Plant Cell Culture" Vol. 4. (D. A. Evans, W. R. Sharp, and P. V. Ammirato, eds.), pp. 457–499. Macmillan, New York.

Becwar, M. R., and Feirer, R. P. (1989). Factors regulating Loblolly pine (*Pinus taeda*) somatic embryo development. *Proc. Southern Forestry Improvement Conf.* 26–28 June, pp. 1–8. North Carolina.

Becwar, M. R., Wann, S. R., and Noland, T. L. (1987). Somatic embryo development and plant regeneration from embryogenic Norway spruce callus. *Tappi J.* **70,** 155–160.

Becwar, M. R., Wann, S. R., Johnson, M. A., Verhagen, V. A., Feirer, R. P., and Nagmani, R. (1988). Development and characterization of *in vitro* embryogenic systems in conifers. *In* "Somatic Cell Genetics of Woody Plants" (M. R. Ahuja, ed.), pp. 1–18. Kluwer Academic Publ., Dordrecht, The Netherlands.

Becwar, M. R., Noland, T. L., and Wyckoff, J. L. (1989). Maturation, germination and conversion of Norway spruce (*Picea abies* L.) somatic embryos to plants. *In Vitro Cell Dev. Biol.* **25,** 575–580.

Becwar, M. R., Wann, S. R., and Nagmani, R. (1989). A survey of initiation frequency of embryogenic callus from immature embryos of ten clones of loblolly pine (*Pinus taeda* L.). *Can. J. For. Res.* **20,** 810–817.

Boulay, M., Gupta, P. K., Krogstrup, P., and Durzan, D. J. (1988). Development of somatic embryos from cell suspension cultures of Norway spruce (*Picea abies* Karst.). *Plant Cell Rep.* **7,** 134–137.

Bourgkard, F., and Favre, J. M. (1988). Somatic embryogenesis from callus of *Sequoia sempervirens*. *Plant Cell. Rep.* **7,** 445–448.

Branch, G. (1987). Pneumatic cylinders drive plant-setting mechanisms. *Agricultural Engineering*. (May/June), p. 16–17.

Buchheim, J. A., Colburn, S. M., and Ranch, J. P. (1989). Maturation of soybean somatic embryos and the transition to plantlet growth. *Plant Physiol.* **89,** 768–775.

Chitwood, D. (1989). Transplanters in transition. *Greenhouse Grower.* (November), p. 18–20.

Dogra, P. D. (1983). Reproductive biology of conifers and its application in forestry and forest genetics. *Phytomorphology* **33,** 142–156.

Dunstan, D. I., Bekkaoui, F., Pilon, M., Fowke, L. C., and Abrams, S. R. (1988). Effects of abscisic acid and analogues on the maturation of white spruce (*Picea glauca*) somatic embryos. *Plant Sci.* **58,** 77–84.

Durzan, D. J. (1985). Tissue Culture Improvement of Woody Perennials: An Overview. In Plant Tissue Culture: Application to Agriculture and Forestry. *In* "Symp. on Plant Cell and Tissue Culture," (September 9–14, 1984), pp. 233–256. University of Tennessee, Knoxville.

Durzan, D. J., and Gupta, P. K. (1987). Somatic embryogenesis and polyembryogenesis in Douglas-fir cell suspension cultures. *Plant Sci.* **52,** 229–235.

Durzan, D. J., and Gupta, P. K. (1988). Somatic embryogenesis and polyembryogenesis in conifers. *Adv. Biotech. Process* **9,** 53–81.

Finer, J. J., Kriebel, H., and Becwar, M. R. (1989). Initiation of embryogenic callus and suspension cultures of white pine (*Pinus strobus* L.) *Plant Cell. Rep.* **8,** 203–206.

Galerine, M., and Dereudder, J. (1987). Survie De Cals Embryogenes D'epice'a Apres Congelation a-196°C. *AFOCEL.* **174.7,** 8–31.

Grand d'Esnon, A., Chee, R., Harrell, R., and Cantliffe, D. (1989). Qualitative and quantitative evaluation of liquid tissue cultures by artificial vision. *Le Technoscope de Biofutur.* **26,** 3–9.

Gray, D. (1981). Fluid drilling of vegetable seeds. *Hort. Rev.* **3,** 1–27.

Gupta, P. K., and Durzan, D. J. (1985). Shoot multiplication from mature trees of Douglas-fir (*Pseudotsuga menziesii*) and sugar pine (*Pinus lambertiana*). *Plant Cell Rep.* **4,** 177–179.

Gupta, P. K., and Durzan, D. J. (1986). Somatic polyembryogenesis from callus of mature sugar pine embryos. *Bio/Tech* **4,** 643–645.

Gupta, P. K., and Durzan, D. J. (1986). Plantlet regeneration via somatic embryogenesis from subcultured callus of mature embryos of *Picea abies* (Norway spruce). *In Vitro Cell Dev. Biol.* **22,** 685–688.

Gupta, P. K., and Durzan, D. J. (1987). Biotechnology of somatic polyembryogenesis and plantlet regeneration in loblolly pine. *Bio/Tech* **5,** 147–151.

Gupta, P. K., Durzan, D. J., and Finkle, B. J. (1987). Somatic polyembryogenesis in embryogenic cell masses of *Picea abies* (Norway spruce) and *Pinus taeda* (loblolly pine) after thawing from liquid nitrogen. *Can. J. For. Res.* **17,** 1130–1133.

Gupta, P. K., Pullman, G. S., Kreitinger, M. E., and Timmis, R. (1989). Somatic embryogenesis in conifers. *Abst. In Vitro Cellular and Development Biology.* **25(3),** 12A.

Gupta, P. K., and Durzan, D. J. (1990). Loblolly pine. *In* Biotechnology of Agriculture and Forestry, Vol. 16 (Y. P. S. Bajaj, ed.), Springer Verlag, Berlin. pp. 383–407.

Haissig, B. E., Nelson, N. D., and Kidd, G. H. (1987). Trends in the use of tissue culture in forest improvement. *Bio/Technology* **5,** 52–59.

Hakman, I., and von Arnold, S. (1985). Plantlet regeneration through somatic embryogenesis in *Picea abies* (Norway spruce). *J. Plant Physiol.* **121,** 149–158.

Hakman, I., Fowke, L. C., von Arnold, S., and Eriksson, T. (1985). The development of somatic embryos in tissue cultures initiated from immature embryos of *Picea abies* (Norway spruce). *Plant Sci.* **38,** 53–59.

Hakman, I., and Fowke, L. C. (1987). An embryogenic cell suspension culture of *Picea glauca* (white spruce). *Plant Cell Rep.* **6,** 20–22.

Hakman, I., and Fowke, L. C. (1987). Somatic embryogenesis in *Picea glauca* (white spruce) and *P. mariana* (black spruce). *Can. J. Bot.* **65,** 656–659.

Hakman, I., and von Arnold, S. (1988). Somatic embryogenesis and plant regeneration from suspension cultures of *Picea glauca* (white spruce). *Physiol. Plantarum* **72,** 579–587.

Hühn, M. (1986). Theoretical studies on the necessary numbers of components in mixtures. 3. Number of components and risk considerations. *Theor. Appl. Genet.* **71,** 622–630.

Jain, S. M., Newton, R. J., and Solta, B. J. (1988). Enhancement of somatic embryogenesis in Norway spruce (*Picea abies*). *Theor. Appl. Genet.* **76,** 501–506.

Jain, S. M., Dong, N., and Newton, R. J. (1989). Somatic embryogenesis in slash pine (*Pinies elliottii*) from immature embryos. *Plant Sci.* **65,** 233–241.

Kartha, K. K., Fowke, L. C., Leung, N. L., Caswell, K. L., and Hakman, I. (1988). Induction of somatic embryos and plantlets from cryopreserved cell cultures of white spruce (*Picea glauca*). *J. Plant Physiol.* **132,** 529–539.

Kermode, A. R., Oishi, M. Y., and Bewley, J. D. (1989). Regulatory roles for abscisic acid in seed developments: a comparison of the evidence from whole seeds and isolated embryos. *In* "Seed Moisture." (P. C. Stanwood and M. B. McDonald, eds.), pp. 23–50. Crop Sci. Soc. American Inc. VSA

Klimaszewska, K. (1989). Somatic embryogenesis and plantlet regeneration from immature zygotic embryos of hybrid larch. *Plant Sci.* **63,** 95–103.

Krogstrup, P. (1986). Embryolike structures from cotyledons and ripe embryos of Norway spruce (*Picea abies*). *Can. J. For. Res.* **16,** 664–668.

Krogstrup, P., Eriksen, E. N., Moller, J. D., and Roulund, H. (1988). Somatic embryogenesis in Sitka spruce (*Picea sitchensis* (Bong.) Carr.). *Plant Cell Rep.* **7,** 594–597.

Lelu, M. A., Boulay, M., and Arnaud, Y. (1987). Formation of embryogenic calli from cotyledons of *Picea abies* (L.) Karst. collected from 3 to 7 days old seedlings. *C. R. Acad. Sci. (Paris)* **305,** 105–107.

Levin, R., Gaba, V., Tal, B., Hirsch, S., DeNola, D., and Vasil, I. K. (1988). Automated plant tissue culture for mass propagation. *Bio/Technology.* **6,** 1035–1040.

Lomax, K., Rodriquez, D., and Kitto, S. (1988). Carrot embryo separation by screening. *ASAE Paper* **#88,** 1025.

LoSchiavo, F. (1984). A critical review of the procedures for embryo purification. *Plant Molecular Biology Reporter* **2,** Number 3, 15–18.

Lu, C. Y., and Thorpe, T. A. (1987). Somatic embryogenesis and plantlet regeneration in cultured immature embryos of *Picea glauca. J. Plant Physiol.* **128,** 297–302.

Lutz, J., Wong, J., Rowe, J., Tricoli, D., and Lawrence, R. (1985). Somatic embryogenesis for mass cloning of crop plant. *Basic Life Sci.* **32,** 105–116.

Nagmani, R., and Bonga, J. M. (1985). Embryogenesis in subcultured callus of *Larix decidua. Can. J. For. Res.* **15,** 1088–1091.

Nagmani, R., Becwar, M. R., and Wann, S. R. (1987). Single-cell origin and development of somatic embryos in *Picea abies* (L.) Karst. (Norway spruce) and *P. glauca* (Moench) Voss (white spruce). *Plant Cell Rep.* **6,** 157–159.

Nagmani, R., and Becwar, M. R. (1988). Factors affecting somatic embryo development in loblolly pine (*Pinus taeda*). *Abst. Proc. 4th Intl. Conifer Tissue Culture working group* (8–12 Aug.), p. 18. Saskatoon, Canada.

Redenbaugh, K., Viss, P., Slade, D., and Fujii, J. A. (1987). Scale up: Artificial seeds. *In* "Plant tissue and Cell Culture" (C. E. Green, D. A. Sommers, W. P. Hackett, and D. D. Biesboer, eds.), pp. 473–493. Alan R. Liss Inc., New York.

Roberts, D. R., Flinn, B. S., and Sulton, B. C. S. (1989). Desiccation promotes and synchronizes the germination of white spruce somatic embryos. I.U.F.R. Working Party on Somatic Cell Genetics, Placerville, California, and NATO Advanced Res. Workshop on Woody Plant Biotechnology, Oct. 15–19, p. 17.

Schuller, A., Reuther, G., and Geier, T. (1988). Somatic embryogenesis from seed explants of *Abies alba. Plant Cell. Tissue Organ Cult.* **17,** 53–58.

Singh, H. (1978). Embryology of Gymnosperms. Encyclopedia of Plant Physiology. Vol. **10,** Part **2,** Gebiuder, Bountraeger, Berlin.

Smith, D. R., Jones, C., and Grace, L. (1989). Somatic embryogenesis in *Pinus radiata:* progress towards artificial seed. *Can. J. For. Res.* (in press).

Thomoson, T. A., Lester, D. T., and Martin, J. A. (1987). Marginal analysis and cost effectiveness in seed orchard management. *Can. J. For. Res.* **17,** 510–515.

Tremblay, F. M. (1988). Somatic embryogenesis and plantlet regeneration from embryos

isolated from stored seeds of *Picea glauca. Abst. In Proc. 4th Intl. Conifer Tissue Culture working group* (8–12 Aug.), p. 20. Saskatoon, Canada.
Verhagen, S. A., and Wann, S. R. (1989). Norway spruce somatic embryogenesis: high-frequency initiation from light-cultivated mature embryos. *Plant Cell. Tissue Organ Cult.* **16,** 103–111.
von Arnold, S., and Hakman, I. (1986). Effect of sucrose on initiation of embryogenic callus cultures from mature zygotic embryos of *Picea abies* (L.) Karst (Norway spruce). *J. Plant Physiol.* **122,** 261–265.
von Arnold, S. (1987). Improved efficiency of somatic embryogenesis in mature embryos of *Picea abies. J. Plant Physiol.* **128,** 233–244.
von Arnold, S., and Woodward, S. (1988). Organogenesis and embryogenesis in mature zygotic embryos of *Picea sitchensis. Tree Physiol.* **4,** 291–300.
von Arnold, S., and Hakman, I. (1988). Regulation of somatic embryo development in *Picea abies* by abscisic acid (ABA). *J. Plant Physiol.* **132,** 164–169.
Warren, G., and Fowler, M. (1977). A physical method for the separation of various stages in embryogenesis of carrot cell cultures. *Plant Sci. Lett.* **9,** 71–76.
Webb, D. T., Webster, F., Flinn, B. S., Roberts, D. R., and Ellis, D. E. (1989). Factors influencing the induction of embryogenic and caulogenic callus from embryos of *Picea glauca* and *Picea engelmannii. Can. J. For. Res.* **19,** 1303–1308.
Webster, S., and Roberts, D. R. (1989). Optimization of somatic embryogenesis and germination of interior spruce for clonal propagation. *Abst. I.V.R.F.O. Workshop on somatic cell genetics.* (Oct. 15–19), p. 29. Placerville, California.
Zobel, B. J., Ikemori, Y., and Campinhos, E. (1985). Vegetative propagation in *Eucalyptus. In* "Clonal Forestry: Its Impact on Tree Improvement and our Future Forests." pp. 136–144. Proc. 19th Meet. (Part 2) Canad. Tree Impr. Ass., August 1983, Toronto.

CHAPTER **5**

Automated Propagation of Microtubers of Potato

Brent H. McCown
Peter J. Joyce

Department of Horticulture
University of Wisconsin-Madison
Madison, Wisconsin

I. INTRODUCTION

The potato (*Solanum tuberosum,* other *Solanum* species, and hybrids) is the world's fourth most important food crop, generating more than $13 billion annually. It ranks behind only rice, wheat, and corn. However, it differs markedly from these crops in that potatoes are clonally (asexually) propagated. Thus the potato is the world's most economically important food crop that is vegetatively propagated. The fact that it is vegetatively propagated, however, has been one of its greatest disadvantages. As the crop is propagated, so are all the tissue-borne viruses and bacteria that have infected the crop in previous seasons. For potato, such pests can lead to significant loss of yields and tuber quality (Burton, 1966). To break this disease cycle, *in vitro* culture of potatoes, including meristem culture/heat treatment and micropropagation, has been practiced for decades. Plants produced from such *in vitro* systems form the nuclear stock for future generations of potatoes. In order to increase the number of tubers and decrease the cost, generations (usually 5–8) of potatoes are grown in fields. Such potatoes are usually termed "seed potatoes."

The above practices have serviced the potato industry well for decades. However, two pressures are creating a demand for new sources of potato propagules. One is the realization that for every generation of field production of seed potatoes, the potential of infection with viruses and bacteria becomes greater; this in turn may lead to future decreases in yield potential (Table I). Thus, there is an increasing demand for "low generation" seed potatoes (ideally no more than four field generations), but still at economically acceptable costs. A second pressure is the widening interest in potatoes as a major food crop around the world (Van der Zaag, 1987; Van Uyen and Van der Zaag, 1985). With the production of potatoes in developing countries where vector-free growing programs may not exist, maintaining a virus-free planting stock is even more difficult. In such cases, partial solutions can be the importing of quality seed (imposing a major cost to the grower), developing small decentralized microculture[1] facilities (Van Uyen and Van der Zaag, 1985; Wattimena, 1983), or utilizing true potato seed (Umaerus, 1987).

Microculture has the potential to be a solution to the shortage of high quality, low generation seed potatoes. The more of the multiplication that is done under the disease-free environments of microculture laboratories, the fewer generations of field planting will be needed to supply the necessary quantities of high quality seed potatoes. However, this potential will be realized only after the major disadvantages of the current *in vitro* methodologies for potatoes are overcome.

Two alternative propagule-types of potatoes can be produced from microculture: transplants derived from shoot cultures and microtubers. Transplants cannot be stored, are sensitive to stress, and require planting conditions and equipment not normally utilized in potato produc-

[1]There has been considerable confusion as to terminology. Our lab (Wattimena *et al.*, 1983; McCown and Wattimena, 1987) and others (Jones, 1988) have used the term *microtuber* to designate potato tubers, regardless of size and whether they are shoot or stolon derived, that are produced entirely *in vitro*. This is consistent with the use of the prefix *micro* for *in vitro* culture (such as microculture, micropropagation, and microcutting). *Minitubers* are defined as tubers produced *ex vitro* (although the plants producing the minitubers may be derived by micropropagation), usually in greenhouses or other horticultural enclosures. Such terminology is consistent with most commercial practices in the United States.

TABLE I

Average Yield by Generation from 1985 through 1989 of Seed Potatoes Grown in Florida by the Maine Seed Potato Certification Program and Incidence of Virus Infection (Mosaic and Leaf Roll) in the 1989 Sample

Generation planted	Yield (cwt/acre)	Total virus (% samples)
N4	280	0.28
G1	295	0.35
G2	275	0.51
G3	269	0.44
G4	222	0.60

Source: Hammond, 1990a, b.

tion. The use of transplants does not integrate well with current potato production practices and thus will always be restricted to specialized growers with the expertise and facilities to handle such sensitive propagules. They are not practical, either economically or functionally, for main crop production in most regions. Microtubers, however, can be handled much like normal seed potatoes; they can be stored, readily shipped, and planted either by hand or automated. Thus microtubers do integrate well with current practices, but their cost at present (greater than $0.25 each) puts them well beyond consideration by the average grower.

Microtubers also integrate well with other complementary technologies being concurrently developed. Microtubers can be the starting material for minituber production. Using minituber technologies provides yet another alternative to scale-up of the number of high quality seed potatoes (Jones, 1988). Additionally, since microtuber technologies potentially provide a much more rapid means of multiplying potatoes than the currently slow process of bulking tuber numbers in the field or greenhouse, they may be ideal for introducing new cultivars/genotypes into the production stream in a minimum of time. This may also be particularly useful in testing and evaluation programs, particularly those associated with the genetic improvement of potatoes using biotechnology where very few initial clonal individuals may exist before greenhouse or field testing.

The only approach to realizing the full potential of microtubers is to drastically reduce the cost of production. Since in most microculture operations, labor accounts for more than 50% of the cost of production, the elimination of the majority of labor input in microtuber production is critical. Two logical solutions to this goal are the use of robotics and the development and commercialization of microtuber bioreactor technology. This paper will emphasize the latter, although many of the concepts will still be applicable to robotic systems.

II. FACTORS TO BE CONSIDERED IN MICROTUBER BIOREACTOR DESIGN

A. Requirements of Any Successful Microtuber Bioreactor

To realize the full potential of microtuber technology, a microtuber bioreactor will probably have to meet at least three requirements: (1) microtuber quantity, (2) microtuber quality, and (3) cost effectiveness.

1. Microtuber Quantity: Be Able to Produce Millions of Microtubers Annually

Microtubers would be planted at the rate of 40,000 to 60,000 propagules per hectare. Thus, in order to make any significant impact on the potato industry, production of very large numbers of microtubers is critical. One factor that helps immensely here is storage. Since microtubers have dormancies like field-produced tubers, they can be stored for months without significant deterioration (Estrada *et al.*, 1986). Thus a production facility can accumulate large numbers of microtubers for a particular planting season and still maintain a uniform output year-round.

However, an obligatory dormancy of several months, which is typical for most potato genotypes, inhibits a very rapid scale-up of production in a minimum amount of time. Thus it would be convenient to have methodologies to artificially break the dormancy of microtubers. Although various methods including hormone treatments (Bruinsma and Swart, 1970; Hemberg, 1970; Turnbull and Hanke, 1985) are available for breaking tuber dormancy, the usefulness with microtubers and the impact on subsequent plant quality and productivity has not been fully documented.

An underlying principle relating to the production of microtubers is that the process should be applicable to a large range of cultivars without major adjustment in the bioreactor design or operation. Many genotypes of potatoes have shown capacities to tuberize *in vitro;* however, the productivity can vary widely (Estrada *et al.*, 1986). Thus it appears that a common bioreactor can be designed that will be applicable for most cultivars; however, the efficiencies may change considerably depending on what particular cultivar is being grown.

2. Microtuber Quality: Produce Microtubers of High and Reproducible Quality

Major factors involved in microtuber quality include uniformity in growth after planting, genetic fidelity, and freedom from disease. Uniform early growth (emergence) of plants derived from microtubers is primarily a function of biological vigor and uniform breakage of tuber dormancy. The size of the microtubers is a concern to field growers because it may be a factor leading to poor vigor in emergence. Test plots have indicated that, given ideal planting and early growth conditions (irrigated light soils), field-planted microtubers can perform as well as regular seed potatoes (McCown and Wattimena, 1987; Wattimena *et al.*, 1983). However, if the field is stressed by wind, drought, frost, herbicide carryover, or insect pressure, microtubers may be at a distinct disadvantage because only one shoot develops from each microtuber and the limited amount of food reserve in the microtuber may not be able to support continued resprouting.

Nonuniform microtuber dormancy has been a problem that has plagued most microtuber production systems. If the time of tuberization *in vitro* varies and/or the growth rates of the individual microtubers differ, the population of microtubers harvested at the end of a particular production cycle can vary markedly in dormancy characteristics; some may be able to sprout immediately while others have a full dormancy requirement. A solution to this problem is to ensure a uniform tuberization time coupled with grading of harvested microtubers to eliminate undersized and thus immature microtubers. Again, an effective artificial method of breaking dormancy may also solve this problem.

Other indicators of tuber quality should also be utilized. Some of these are apparent, as is the assurance of freedom of viral, mycoplasma, and bacterial diseases and the guarantee of genetic trueness-to-type. The first is readily accomplished by the use of established biological and molecular testing procedures (Johansen *et al.*, 1984; Wright, 1988); the second characteristic is much more difficult to monitor. Because of the common incidence of somaclonal variation in potatoes generated from *in vitro* adventitious systems (Karp *et al.*, 1989), the meristems generating the microtubers will necessarily have to be from preformed axillary buds or stem/stolon tips to promote genetic stability. The issue of how to verify that genetic change has not occurred in the bioreactor is diffi-

TABLE II
The Percent Dry Matter and Percent Calcium in Microtubers of Potato cv. Red Pontiac Tuberized with and without Hormones in the Medium

Treatment	Percent dry matter	Percent calcium
No hormones	9.4	0.079
Cytokinin only	15.7	0.014
Coumarin only	11.1	0.088
Cytokinin + coumarin	18.4	0.019

Source: Joyce, 1988.

cult to answer. If a subsample of microtubers has to be grown out to maturity to observe genetic fidelity (a process that will take months), this will markedly reduce the flexibility of the microtuber production system. At the very least, any firm selling microtubers should have a well-coordinated field evaluation program to monitor the performance of its microtubers under grower conditions (Johanson *et al.*, 1989).

Other biological quality factors are much less obvious. For example, calcium nutrition has been increasingly pinpointed as important in tuber storage problems (McGuire and Kelman, 1984) and in potato growth and yield (Dyson and Digby, 1975; Palta, 1989). The method of inducing tuberization of microtubers can have a marked effect on calcium content of the resultant microtubers (Table II); whether such differences will affect microtuber performance, either in storage or emergence, is unknown.

3. Cost Effectiveness: Produce Microtubers at a Cost Competitive with Other Accepted Cloning Methods

As the costs are lowered, more markets can be accessed. At the high range, microtubers will be competitive only with minitubers or nuclear (prebasic; elite) seed which is used to start the standard seed production cycle. At an intermediate level of cost, microtubers will compete with seed produced for several generations in fields (basic seed). At a low cost, probably approaching one cent apiece, microtubers will compete with normal certified seed.

B. Factors Controlling Microtuber Production

A general flowchart of a microtuber production process will generally have a number of steps:

I. Source tissue
|
II. Shoot culture stabilization
|
III. Multiplication of shoot meristems
|
IV. Tuberization of meristems
|
V. Microtuber growth and maturation
|
VI. Harvest
|
VII. Storage/shipping
|
VIII. Planting

Since potatoes are readily microcultured, the first two steps are rather routine and uncomplicated except for the essential step of producing tissue that is free of known viral and bacterial pathogens. The reader is directed to other reviews on this topic (Wang, 1977; Wright, 1988).

The third step, multiplication of meristems, is, surprisingly, not as readily achieved with potatoes as with many herbaceous crops. This step is critical since it is the preformed meristems that are the ultimate source of the microtubers. Biologically, every meristem, whether nodal or apical, is capable of tuberizing under inductive conditions. Thus the basis of microtuber multiplication is first the multiplication of stems (either shoots or stolons) which will subsequently provide the meristems for tuberization. However, two major impediments are evident: (1) potato shoot cultures do not respond well to the classical cytokinin stimulation of axillary shoot development observed in the micropropagation of most crops, and (2) during the tuberization process, there is such strong competition between meristems along a stem axis that the first meristem to tuberize inhibits the tuberization of the other meristems.

Table III shows a typical cytokinin response curve of potato shoot explants using benzyladenine (BA). Even at concentrations (above 4 μM BA) which show toxicity, shoot multiplication is not stimulated above that of the controls. This is the case with other cytokinins as well. Thus, the multiplication of meristems through the cytokinin stimulation of axillary bands is not practical. A common solution to this problem has been to manually sever a stem into nodal segments, each of which will produce one shoot. However, such an approach will be very labor intensive or dependent on sophisticated robotic-type systems.

TABLE III
Growth/Response Curve for the Cytokinin Benzyladenine and *In Vitro* Shoot Multiplication of the Potato cv. Katahdin[a]

Benzyladenine concentration (μM)	Final total shoots/explant	Final total nodes/explant
0	3.2 (0.4)	15.4 (1.6)
0.4	3.2 (0.4)	12.8 (1.6)
1.0	3.2 (0.4)	13.2 (11.7)
4.0	3.6 (0.5)	14.4 (1.6)
10.0	3.4 (0.5)	14.2 (2.9)
40.0	0.6 (0.8)	2.6 (4.3)

[a]The original explant contained three nodes. Standard error in parentheses (from Joyce, 1988).

The competition between meristems along an axis during tuberization is another important problem. If a typical shoot or stolon axis is placed under tuberizing conditions, tuberization can first occur at almost any node. However, once this occurs, another node will usually not form a tuber (Fig. 1), or if it does, it will be considerably delayed. Even after

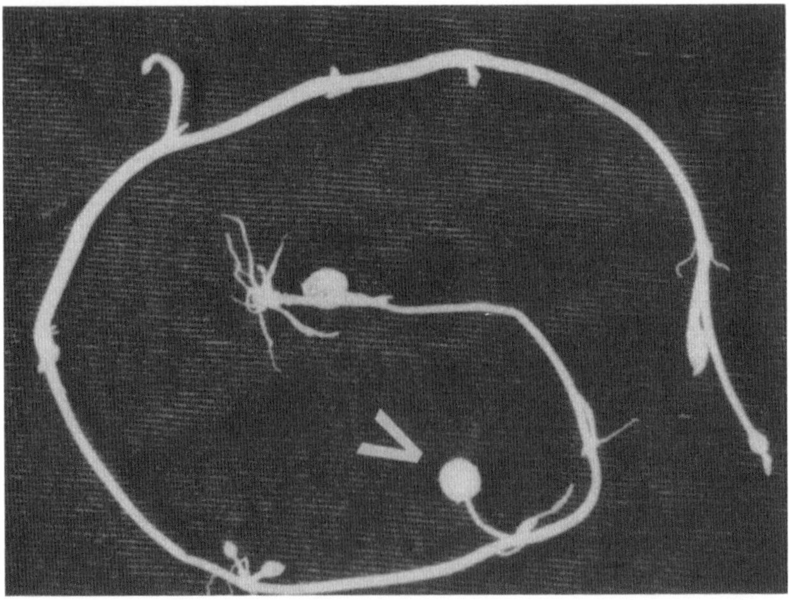

Fig. 1. An etiolated stem from a potato culture grown for 30 days under high tuberizing conditions (8% sucrose, 22 μM cytokinin, 85 μM coumarin, 15°C, dark). Note that only one usable microtuber (arrow) has formed on the stem axis even though the stem has more than eight nodes.

extensive culture periods, most of the nodes will never tuberize. Because uniform tuber development is largely dependent on uniform time of tuberization, functionally only one commercial microtuber is obtained for every independent shoot or stolon axis cultured. This fact has been the major factor limiting the rapid and low cost multiplication of microtubers. Again, a common solution has been the labor-intensive and tedious process of mechanically dividing the nodes along an axis into isolated single node units, each of which will then tuberize when placed in culture vessels (Fig. 2).

The factors involved in initiating *in vitro* tuberization of cultured meristems are well known (Ewing, 1985; Garcia-Torres and Gomez-Campo, 1973; Hussey and Stacy, 1984; Joyce, 1988; Koda and Okazawa, 1983; Mingo-Castel *et al.*, 1974, 1976a,b; Palmer and Barker, 1973; Roca *et al.*, 1988; Rosell and Bertoldi, 1987; Stallknecht, 1972; Stallknecht and Farnsworth, 1982; Wang and Hu, 1985; Wattimena, 1983). The most critical include cool temperatures (usually 15–25°C), high medium sucrose levels (6–8%), and growth regulators (high cytokinin and/or coumarin and/or growth retardants). The exact conditions will vary with the particular laboratory as there is considerable interaction between factors and thus no one optimum probably exists. In addition, genotype influ-

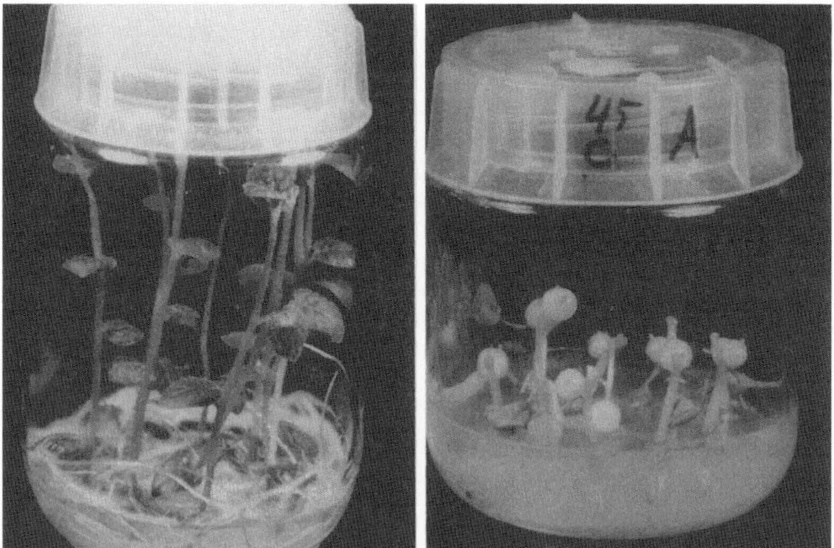

Fig. 2. A commonly used method of producing microtubers. The process begins by establishing shoot cultures (left) from which single nodes are manually isolated. These nodal explants are then subcultured to a tuberizing medium and environment where single microtubers form on each explant (right). The microtubers here have been developing for 8 weeks.

ences, the exact tissue being tuberized, the type of solidifying agent (if any), and the vessel/closure can have strong effects. For example, cytokinins will probably be an obligatory requirement if the tuberization occurs in the dark but are usually not required in lighted conditions (Wattimena, 1983).

The control of uniformity in the time of tuberization is probably the most important economic variable. All of the microtuber production schemes of which we are aware are batch cultures. Thus, a uniform tuberization period will be critical for the harvest of a batch of microtubers of high physiological uniformity. The time of tuberization and its uniformity will be a complex function of both the tissue being cultured and the tuberization stimuli. The uniformity of the tissue being tuberized is obviously one important factor. In most systems, the tuberization is initiated by suddenly changing both the medium and growth environment of a stem/stolon culture that has been growing uniformly in a vegetative state. This may also be coupled with the singulation of individual meristematic (nodal) areas.

Microtubers have a linear growth curve in regard to biomass (Fig. 3) and a sigmoid curve in regard to size (Fig. 4). In most systems, usable microtubers of a minimum of 50 mg in weight and 4 mm diameter will require 6–8 weeks of culture after tuberization begins. In comparison with many other bioreactor systems, this is a rather extended culture period and thus puts a premium on low maintenance/low input designs for control of costs.

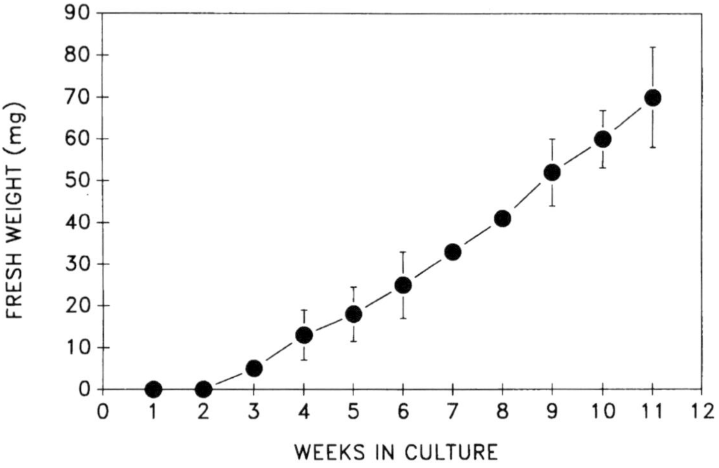

Fig. 3. The increase in fresh weight over time of "Red Pontiac" microtubers grown on 45 μM cytokinin and 170 μM coumarin. Bars indicate one standard deviation. (From Joyce 1988.)

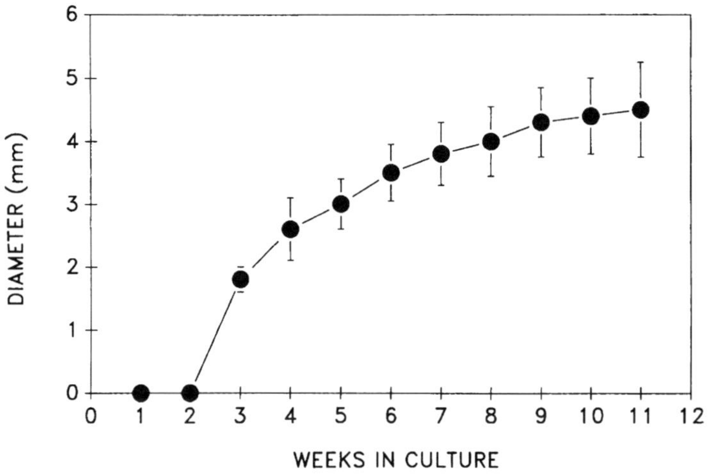

Fig. 4. The increase in diameter over time of "Red Pontiac" microtubers grown under the same conditions as in Fig. 3. (From Joyce 1988.)

Since most microtuber productions systems are batch cultures, harvesting microtubers from a bioreactor is not overly complicated. However, the separation of the microtubers from the parent stem/stolon/root mass may be difficult but should not result in injury to the microtuber skin (which is usually not fully matured at the time of harvest from a bioreactor).

Microtubers have dormant periods equal to or even greater than regular tubers of the same cultivar. Microtubers of good size (50 mg plus) and substance (more than 10% dry matter) will store well in cold (4–10°C) and humid conditions. Enough air circulation to maintain a normal gaseous environment (adequate oxygen, low ethylene) is important if large numbers are stored in common. A period of partial drying just after harvest may help to mature the skin and prevent physiological breakdown problems in storage.

III. A NOVEL MICROTUBER BIOREACTOR

One approach to producing microtubers in mass and with a minimum of labor input is being developed by Small Potatoes, Inc. in Madison, Wisconsin. This microtuber production system is of interest because some unique approaches have been developed to overcome the major problems outlined above. The exact process, termed the "Microtuber

Multiplication System" or MMS is proprietary and is patent-pending; however, some of the basic concepts were developed at the University of Wisconsin-Madison and include the following:

1. The basic bioreactor is a cylinder that is rotated slowly on its horizontal axis (Fig. 5). The shoot/stolon/microtuber complex is held to the outside of the bioreactor during rotation. Liquid medium is added to the bioreactor to a level that fills less than 10% of the volume of the vessel. As the vessel rotates, the tissue is intermittently bathed in medium. At the ends of the bioreactor are vents for controlling gas exchange.

2. The bioreactor is charged with microtubers produced previously. The culture moves through three distinct stages. The first stage of culture (multiplication phase) involves sprouting the microtuber under conditions that promote the development of shoot-tip necrosis (Sha *et al.*, 1985). In this way, apical dominance is broken and axillary bud development is stimulated. These axillary shoots also develop shoot-tip necrosis, and more branching is stimulated. To begin the second stage (elongation phase), a medium stimulatory to normal shoot growth is added and shoot/stolon elongation occurs from the many shoot bases established during the first phase. These stems also root as they grow, creating many independent axes. The third stage (tuberization) is initiated by changing both the environment and the medium. A mass of microtubers produced by a similar system is shown in Fig. 6.

Thus, this design manages to overcome many of the major problems in microtuber production systems: (1) The unresponsiveness of potatoes to cytokinins is avoided by employing shoot-tip necrosis for the stimulation of branching of shoots from microtubers. (2) Labor input is mini-

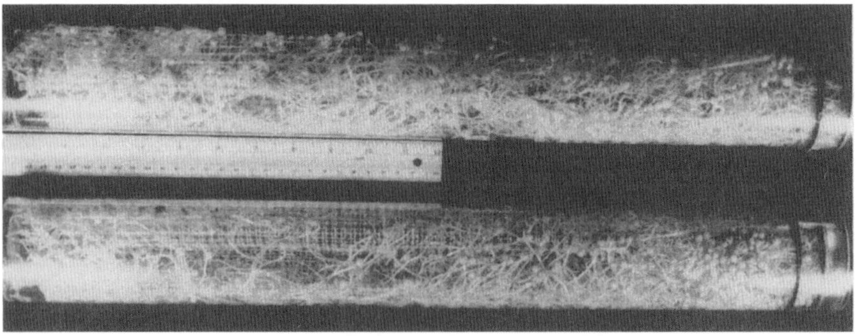

Fig. 5. A linear bioreactor in which potato microtubers are being produced. The reactor uses a liquid medium and is rolled slowly to alternately bathe and aerate the explants. This bioreactor is 60 cm long. (Courtesy of Small Potatoes, Inc., Madison, Wisconsin.)

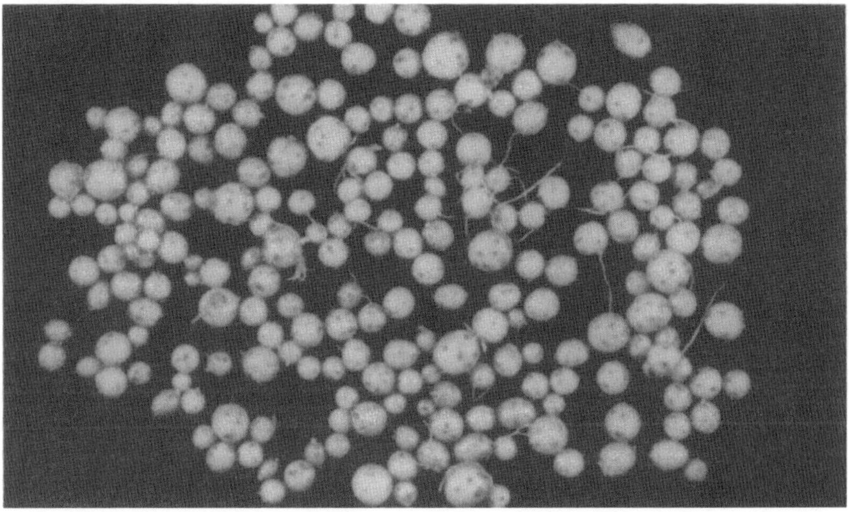

Fig. 6. A sample of the microtubers harvested from a bioreactor such as that in Fig. 5. (Courtesy of Small Potatoes, Inc., Madison, Wisconsin.)

mal. The bioreactor design allows all the manipulations to be done in one vessel without any direct manual manipulations beyond that of the initial inoculation and final harvest. (3) Costs are low. The system is simple and requires a low-cost growth environment.

IV. CONCLUSIONS

The use of microcultured propagules has proven an ideal solution to breaking many of the disease cycles threatening a potato crop. The microtuber appears to be the most logical and practical product that can be produced in such micropropagation systems. However, how millions of microtubers can be produced at competitive costs has not been fully established. Microtuber bioreactors may be one logical approach.

No matter what production systems are finally proven feasible, considering the demand for high-quality potato propagules coupled with the ability to rapidly scale-up new genotypes and coupled with other biotechnological approaches, microtuber-based technologies promise to be an exciting development in the near future.

REFERENCES

Bruinsma, J., and Swart, J. (1970). Estimation of the course of dormancy of potato tubers during growth and storage, with the aid of gibberellic acid. *Potato Res.* **13,** 29–40.

Burton, G. W. (1966). "The Potato." (H. Veenman and NV Zonen, eds.). Wageningen, Holland. 382 p.

Dyson, P. W., and Digby, J. (1975). Effects of calcium on sprout growth and subapical necrosis in Majestic Potatoes. *Potato Res.* **18,** 290–305.

Estrada, R., Tovar, P., and Dodds, J. H. (1986). Induction of *in vitro* tubers in a broad range of potato genotypes. *Plant Cell, Tissue Organ Cult.* **7,** 3–10.

Ewing, E. E. (1985). The role of hormones in potato (*Solanum tuberosum* L.) tuberization. *In* "Potato Physiology" (P. H. Li, ed.), pp. 515–538. Academic Press, Orlando.

Garcia-Torres, L., and Gomez-Campo, C. (1973). *In vitro* tuberization of potato sprouts as affected by ethrel and gibberellic acid. *Potato Res.* **16,** 73–79.

Hammond D. F. (1990a). Yields, by generation, of seed potatoes planted in Maine. *Maine Potato News* **4(1),** 14.

Hammond, D. F. (1990b). An early Florida test accomplished. *Maine Potato News* **4(1), 1,** 6–7.

Hemberg, T. (1970). The action of some cytokinins on the rest-period and the content of acid growth inhibiting substances in potato. *Physiol. Plant.* **23,** 850–858.

Hussey, G., and Stacey, N. J. (1984). Factors affecting the formation of *in vitro* tubers of potato (*Solanum tuberosum* L.). *Ann. Bot.* **53,** 565–578.

Johansen, D. G., Knutson, K. W., Slack, S. R., and Jones, E. D. (1984). Recommendations for standards and guidelines for acceptance of tissue cultured potato seed stocks into certification programs. *Amer. Pot. J.* **61,** 368–370.

Jones, E. D. (1987). Trends in commercial and seed potato production in North America and expected future developments. *In* "Recent and Future Developments of the Potato in the World" *Acta Hortic.* **213,** 105–120.

Jones, E. D. (1988). A current assessment of *in vitro* culture and other rapid multiplication methods in North America and Europe. *Am. Pot. J.* **65,** 209–220.

Joyce, P. J. (1988). Microtuber production as an alternative means of potato propagation. MS Thesis, University of Wisconsin–Madison. 101 p.

Karp, A., Jones, M. G. K., Foulger, D., Fish, N., and Bright, S. W. J. (1989). Variability in potato tissue culture. *Am. Pot. J.* **66,** 669–684.

Koda, Y., and Okazawa, Y. (1983). Influences of environmental, hormonal and nutritional factors on potato tuberization *in vitro*. *Japan J. Crop Sci.* **52,** 582–591.

McCown, B. H., and Wattimena, G. A. (1987). Field performance of micropropagated plants. *In* "Biotechnology in Agriculture and Forestry" (Y. P. S. Bajaj, ed.), Vol. 3, pp. 80–88. Springer-Verlag, Berlin.

McGuire, R. G., and Kelman, A. (1984). Reduced severity of *Erwinia* soft rot in potato tubers with increased calcium content. *Phytopathology* **74,** 1250–1256.

Mingo-Castel, A. M., Negm, F. B., and Smith, O. E. (1974). Effect of carbon dioxide and ethylene on tuberization of isolated potato stolons cultured *in vitro*. *Plant Physiol.* **53,** 798–801.

Mingo-Castel, A. M., Smith, O. E., and Kumamoto, J. (1976a). Studies on the carbon dioxide promotion and ethylene inhibition of tuberization in potato explants cultured *in vitro*. *Plant Physiol.* **57,** 480–485.

Mingo-Castel, A. M., Young, R. E., and Smith, O. E. (1976b). Kinetin-induced tuberization of potato *in vitro:* on the mode of action of kinetin. *Plant Cell Physiol.* **17,** 557–570.

Palmer, C. E., and Barker, W. G. (1973). Influence of ethylene and kinetin on tuberization and enzyme activity in *Solanum tuberosum* L. stolons cultured *in vitro. Ann. Bot.* **37,** 85–93.

Palta, J. P. (1989). Spoonfeeding tubers during bulking: A new concept in potato nutrition. *The Badger Common'tator* **41,** 27.

Roca, W. M., Espinoza, N. O., Roca, M. R., and Bryan, J. E. (1968). A tissue culture method for the rapid propagation of potatoes. *Am. Pot. J.* **55,** 691–701.

Rosell, G., and Bertoldi, F. G. (1987). *In vitro* mass tuberization as a contribution to potato micropropagation. *Potato Res.* **30,** 111–116.

Sha, L., McCown, B. H., and Peterson, L. A. (1985). Occurrence and cause of shoot-tip necrosis in shoot cultures. *HortScience* **110,** 631–634.

Stallknecht, G. F. (1972). Coumarin-induced tuber formation on excised shoots of *Solanum tuberosum* L. cultured *in vitro. Plant Physiol.* **50,** 412–413.

Stallknecht, G. F., Farnsworth, S. (1982). General characteristics of coumarin-induced tuberization of axillary shoots of *Solanum tuberosum* L. cultured *in vitro. Am. Pot. J.* **59,** 17–32.

Turnbull, C. G. N., and Hanke, D. E. (1985). The control of bud dormancy in potato tubers. Evidence for the primary role of cytokinins and a seasonal pattern of changing sensitivity to cytokinin. *Planta* **165,** 359–365.

Umaerus, M. (1987). True potato seed. *Proc. 10th Triennial Conf. EAPR,* pp. 72–102. Aalborg, Denmark.

Van Uyen, N., and Van der Zaag, P. (1985). Potato production using tissue culture in Vietnam; the status after four years. *Am. Pot. J.* **62,** 237–241.

Van der Zaag, P. (1987). Developments in potato production techniques in Asia. *In* "Recent and future developments of the potato in the world." *Acta Hortic.* **213,** 79–80.

Wang, P. J. (1977). Regeneration of virus-free potato from tissue culture. *In* "Plant Tissue Culture and Its Bio-technological Application" (W. Bartz, E. Reinhard, and M. H. Zenk, eds.), pp. 286–391. Springer-Verlag, Berlin.

Wang, P. J., and Hu, C. (1985). Potato tissue culture and its application in agriculture. *In* "Potato Physiology" (P. H. Li, ed.), pp. 504–564. Academic Press, Orlando.

Wattimena, G. A. (1983). Micropropagation as an alternative technology of potato production in Indonesia. PhD thesis, University of Wisconsin, Madison, WI, USA, 201 p.

Wattimena, G. A., McCown, B. H., and Weis, G. (1983). Comparative field performance of potatoes from microculture. *Am. Pot. J.* **60,** 27–33.

Wright, N. S. (1988). Assembly, quality control, and use of a potato cultivar collection rendered virus-free by heat therapy and tissue culture. *Am. Pot. J.* **65,** 181–198.

CHAPTER **6**

Automated Propagation of Microbulbs of Lilies

Shinsaku Takayama

Department of Biological Science and Technology
Tokai University
317 Nishino, Numazu,
410-03 Shizuoka, Japan

Brad Swedlund

Native Plants, Inc.
Salt Lake City, Utah

Yoshiyuki Miwa

Department of Science and Engineering
Waseda University
Tokyo, Japan

I. INTRODUCTION

Lilium species and their hybrids are among the most important ornamental bulbous plants. They are grown worldwide and constitute a large share of the worldwide market of ornamental plants. Since natural clonal propagation is very slow, artificial methods such as bulb splitting, bulb scaling, or stem cutting have traditionally been used for clonal propagation (Hartmann and Kester, 1975). However, these methods are problematic because production of large numbers of genetically homogenous plants is tedious and time consuming. Moreover, deterioration of flower quality due to an increasing virus titer has emerged as a major problem resulting in part from the use of these traditional techniques.

In lilies, *in vitro* culture methods have been employed to obtain virus-free plants (Sheridan, 1968; Allen and Fernald, 1972; Allen, 1974; Allen and Linderman, 1976). Viruses of lilies are transmitted by aphids and, as mentioned, their titer is maintained and even increased by traditional means of vegetative propagation (Allen, 1974; Mowat and Stefanac, 1974). Therefore, once virus-free material has been obtained, it is desirable to keep this material free from viruses as long as possible before growing under commercial greenhouse or field conditions where virus reinfestation is likely to occur. Propagation of virus-free lily plants through *in vitro* culture techniques has been studied by many researchers, including Robb (1957), Riviere (1968), Sheridan (1968), Sharp *et al.* (1971), Simmonds and Cumming (1976), Anderson (1977), Aartrijk and Blom-Barnhoorn (1981), Novak and Petru (1981), Takayama and Misawa (1979, 1980, 1982a), and Liu and Burger (1986). The most common multiplication method used in these reports involves regeneration of bulblets from bulbscale explants. It is possible to produce lily plants rapidly and in large numbers using this basic method.

Use of callus also seems to be an important tool for rapid propagation of lilies (Simmonds and Cumming, 1976; Stimart *et al.*, 1980). Neither phenotypic (Simmonds and Cumming, 1976) nor chromosomal (Bennici, 1976) aberrations have been observed using this technique, and the potential for the production of large numbers of plants from these callus cultures has been suggested.

In practice, mass propagation of a large number of bulbs through *in vitro* culture has been achieved using modified methods of shake- and/or bioreactor-culture techniques (Takayama and Misawa, 1982b,c, 1983). These techniques are amenable to scale-up by automated culture processes. In this report we discuss some of the fundamental concepts and problems of scaled-up production of *Lilium* bulbs, with particular emphasis on development of systems employing regeneration of bulblets from bulbscales. The basic areas of discussion are: (1) Large scale propagation in liquid medium using bioreactors. (2) Large scale regeneration of microbulbs under nonaseptic conditions. (3) Automechanization of the tissue culture processes traditionally accomplished by hand. (4) Speculation about the development of a "plant factory" for large scale, economical production of *Lilium* bulbs.

The procedures outlined here have been directed principally at the production of *Lilium* bulbs. However, the processes have been deliberately designed with a view that other species may benefit by such production schemes as well. In particular, we feel that these procedures are especially amenable to production of other bulbous plants with appropriate modifications of procedures.

II. REGULATION OF *IN VITRO* DIFFERENTIATION AND GROWTH IN *LILIUM* FOR LARGE-SCALE PROPAGATION

A. General Procedures

The process leading to establishment of lily plants *in vitro* can be divided into adventitious and embryoidal plantlet formation. The former process is the direct regeneration of microbulbs from vegetative organs or tissues, especially from the bulbscales (Stimart and Ascher, 1978; Anderson, 1977; Novak and Petru, 1981; Takayama and Misawa, 1979, 1980, 1981, 1982a,b,c, 1983), and the latter process is asexual embryogenesis through callus (Simmonds and Cumming, 1976; Stimart *et al.*, 1980; Takayama and Misawa, 1983). The former was selected as a suitable method for clonal propagation in *Lilium* for several reasons:

1. The period for regeneration of plants from callus is much longer than that of adventitious plantlet formation from bulbscales.
2. Regeneration of plantlets from callus results in the termination of growth of callus, and conversely, active growth of callus results in virtually no plantlet formation (Fig. 1) (S. Takayama and M. Misawa, unpublished results).

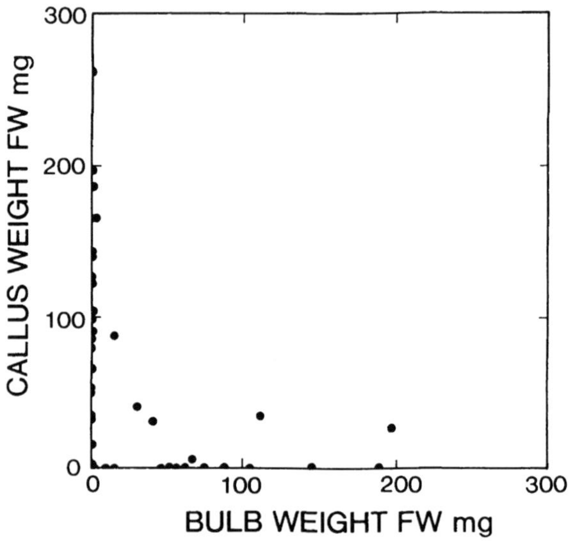

Fig. 1. Relationship between growth of bulbs and growth of callus in *Lilium longiflorum* bulbscales cultured *in vitro*. n, 40.

3. Although lily plants regenerated from callus have been shown to be phenotypically homogenous (Simmonds and Cumming, 1976), genetic aberrations in plants regenerated from callus of other species is widely known (D'Amato, 1977). This can present a problem in commercial, large-scale production since even a seemingly low percentage of aberrant plants can significantly affect the costs of production. Therefore, it is desirable to develop a system which is applicable to other plant species which may show genetic aberrations upon regeneration from callus but do not show aberrations from propagation by axillary shoot formation or direct organogenesis.
4. Bulblets regenerated from bulbscales are easily maintained *in vitro*. Bulbscales from these *in vitro*–produced bulblets are highly efficient in new bulblet formation and are the best material for propagation of *Lilium* bulblets *in vitro*.

B. Phytohormones

Factors affecting differentiation and growth of *Lilium in vitro* have been studied extensively. Many studies on regulation of morphogenesis using phytohormones have been carried out to elucidate the differentiation phenomena, and most of the results support the observation by Skoog and Miller (1957) that the balance of phytohormones is important. Morphogenesis consists of two processes, i.e., "differentiation" and "growth."

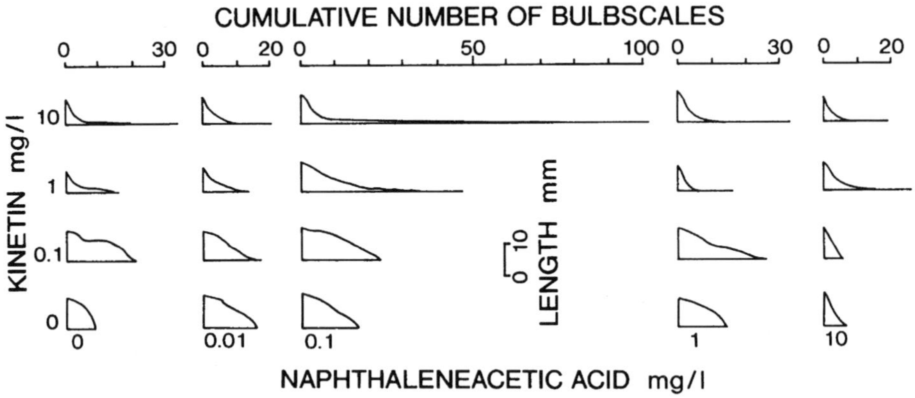

Fig. 2. Effect of NAA and kinetin on the neoformation of bulbscales in *Lilium auratum* cultured *in vitro* (Takayama and Misawa, 1982a).

From this point of view, active differentiation does not necessarily lead to active growth. For example, higher kinetin concentrations stimulate bulbscale differentiation but inhibit bulb growth in *Lilium* (Fig. 2). The addition of 0.1 mg/l naphthaleneacetic acid (NAA) to the medium synergistically stimulates kinetin activity in differentiating bulbscales. The specific cytokinin employed is important as well. A comparison of the effects of benzyladenine (BA) and kinetin are shown in Fig. 3. The two cytokinins are specific for the plant species, but, in general, BA has a

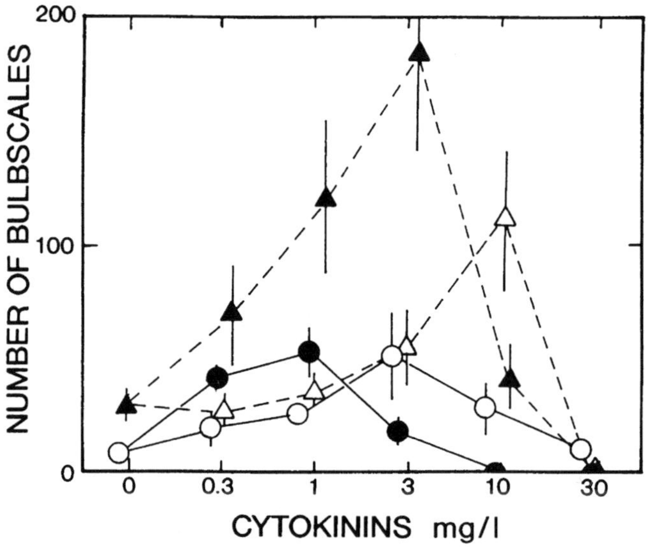

Fig. 3. Effect of BA and kinetin on differentiation of bulbscales of *Lilium auratum* and *L. speciosum* in tissue culture. Symbols: *L. auratum* (triangle), *L. speciosum* (circle), BA (filled), Kinetin (open) (Takayama and Misawa, 1982a).

stronger stimulative effect than kinetin on bulbscale differentiation but also has a stronger effect on the inhibition of growth (Takayama and Misawa, 1982a).

C. Sucrose and Salt Concentrations

In addition to the effects of the phytohormones, growth and differentiation are also affected by the sucrose and salt concentrations of the medium (Takayama and Misawa, 1981, 1982b). In *Lilium*, increasing the sucrose concentration in the medium from 3% to 9% or 12% has resulted in an increase in the size of microbulbs (Niimi and Onozawa, 1979; van Aartrijk and Blom-Barnhoorn, 1980; Takayama *et al.*, 1982). A decrease in the multiplication rate, along with a decrease in scale leaf formation, has been noted when bulb enlargement was induced by higher sucrose concentrations.

As previously reported (Takayama and Misawa, 1982a), higher concentrations of sucrose appear to have an inhibitory effect on kinetin activity (Fig. 4). Multiplication rates diminish markedly with higher concentrations of sucrose and without a change in cytokinin levels. When kinetin levels are increased along with sucrose concentrations, multiplication continues at a high rate, but bulblet enlargement is minimal compared with growth on high sucrose without an increase in kinetin. Higher sucrose concentrations also resulted in increases in root differentiation and growth. Higher salt concentrations inhibited this root development.

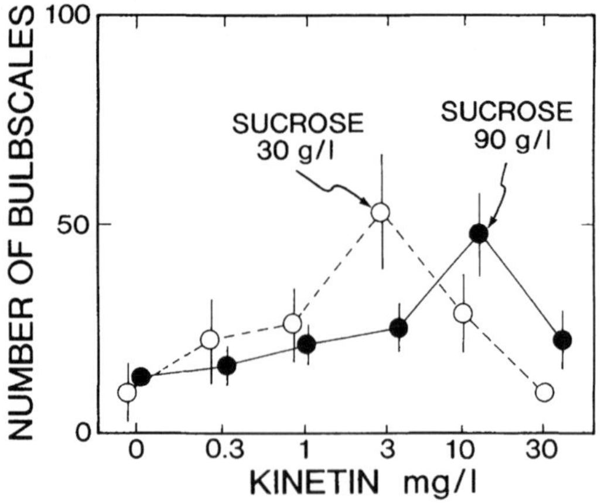

Fig. 4. Interaction of sucrose and kinetin on differentiation of bulbscales of *Lilium speciosum* in tissue culture (Takayama and Misawa, 1982a).

It is likely that there is an osmotic component to the effect of sucrose on bulblets of lilies. To some extent the response can be reproduced by substituting mannitol for part of the sucrose in a high sucrose medium, but the effect is not as pronounced as with sucrose alone (data not shown). The effects associated with sucrose in relation to kinetin would also suggest that the effect is not entirely an osmotic one. One possible mechanism may be related to the concentration effects of medium salts or sucrose on cytokinin uptake by the plant tissue. This may be related to the hydrophobicity of cytokinin, which has been indicated by the partitioning of cytokinin between organic solvent and water (Hansch *et al.*, 1963; Hansch and Fujita, 1964).

The concentration of the salts in the medium also affects the morphology and growth of the cultures (Takayama and Misawa, 1979). Some of the effects of MS (Murashige and Skoog, 1962) salt concentration on growth in bioreactors are discussed below (Section IIIB).

III. AUTOMATED PROPAGATION OF MICROBULBS FROM BULBSCALES USING BIOREACTOR TECHNIQUES

A. The Fundamental Process

The process is essentially the same for several lily species, with some variation for optimal medium components (Takayama and Misawa, 1982b,c, 1983). The general scheme employed is presented in Fig. 5 and outlined in more detail below. Bulbs of stock plants grown in the soil (A) are sterilized with sodium hypochlorite and ethanol solutions (B), then sectioned into small pieces of about 0.5–1.0 cm^2 (C). Each piece is transferred into a 25 × 125 mm test tube containing MS medium supplemented with 30–90 g/l sucrose, 0.1 mg/l NAA, and 8 g/l agar. Cultures are placed under 2.5 $W \cdot m^{-2}$ of continuous irradiance at 25°C (D, E). Bulbscales (G) dissected from a bulblet grown *in vitro* (F) are transferred to MS agar medium containing high concentrations of cytokinins (3–10 mg/l kinetin, BA, or 4PU) (H), where numerous primordialike adventitious bulbscales are regenerated (I). The bulbscales are grown rapidly in shake culture (J, K) and yield numerous secondary bulbscales (L). These can be used as the material for bulblet formation on agar or in liquid medium using a shake flask (M, N) or a bioreactor (O, P). Bulblets thus produced (Q) are treated at 5°C to break dormancy and then transplanted to the soil (R). Treated bulblets produce shoots (S) and flowers (T). The use of the liquid shake culture method or the bioreactor culture

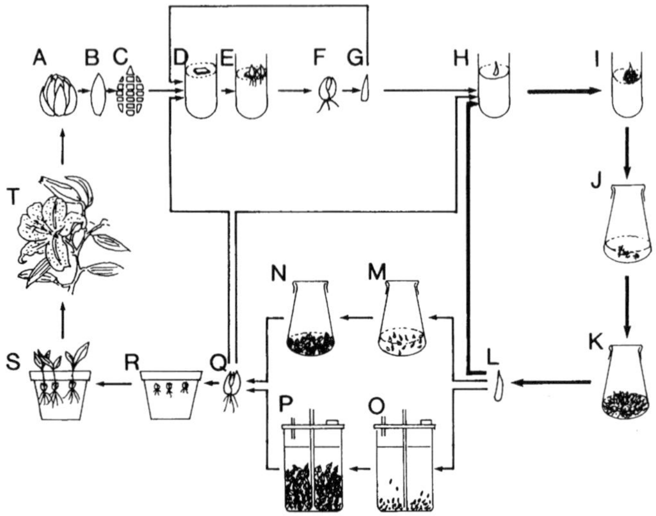

Fig. 5. Schematic representation of rapid clonal propagation of *Lilium* by shake- or bioreactor-culture (Takayama and Misawa, 1982b, partly modified).

method in this scheme (J to K and O to P) is suitable for a mass propagation system. The number of the bulbscales can be increased rapidly by recycling the procedures in H through L in this figure. The bioreactor portion of the scheme is described in more detail below.

B. Medium Conditions for Successful Regeneration and Growth of *Lilium* Microbulbs in a Bioreactor

Although the medium composition for bioreactor cultures is fundamentally the same as that for agar cultures, the effect of the composition of the medium on microbulb formation is more exacting in a bioreactor. The strength of MS salts affects bulblet formation of *Lilium longiflorum*. A half ($\frac{1}{2}\times$) strength MS medium stimulated bulblet formation, but normal ($1\times$) and double ($2\times$) strength MS medium repressed bulblet formation and stimulated callus formation (Fig. 6). The principal medium component involved in this phenomenon is ammonium ion concentration from ammonium nitrate (S. Takayama, unpublished results). Reduction or deletion of ammonium nitrate from the medium stimulated microbulb regeneration and completely inhibited callus formation. Addition of abscisic acid (ABA) in the concentration range from 0.1 to 0.3 mg/l also stimulated microbulb regeneration and inhibited scale leaf and callus formation (Takayama *et al.*, 1986). Although these are typical characteristics of *L. longiflorum,* similar phenomena were observed in other *Lilium* species as well. Half strength MS medium or revised MS medium (ammonium nitrate reduced or eliminated) in combination with ABA is suitable for microbulb formation in a bioreactor.

The size distribution of microbulbs of *L. longiflorum* propagated in a 2-liter glass bioreactor was investigated. A total of 634 microbulbs were produced from 250 bulbscales within 2 months (Fig. 7). About 70% of these exceeded 0.3 grams (fresh weight) and flowered during the first year of cultivation in the soil. This is a substantial increase in flowering when compared with a previous report in which about 40% of *in vitro*–grown *L. longiflorum* bulblets flowered within 1 year (Takayama *et al.*, 1982).

Sucrose concentration is another important factor regulating microbulb regeneration and growth. The number of microbulbs produced per bulbscale was highest in the presence of 1–3% sucrose, but the enlargement of microbulbs was highest at 6–9% sucrose.

It is clear that other species and hybrids may take longer to flower than *L. longiflorum,* whether they are produced by traditional methods

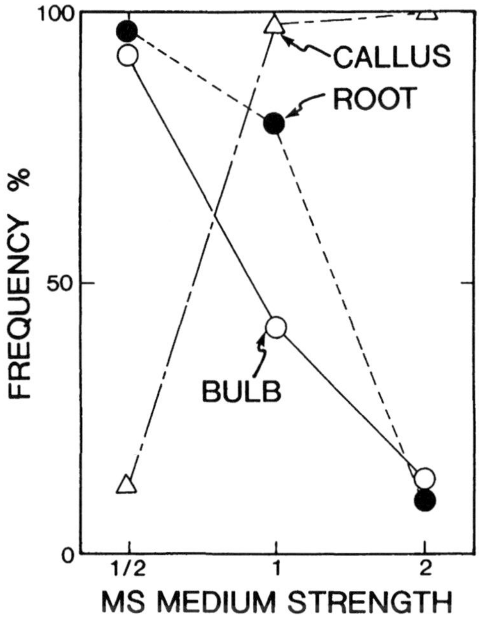

Fig. 6. Effect of MS medium strength on bulblet, root, and callus formation in *Lilium longiflorum* in a fermentor (Takayama and Fukano, in preparation).

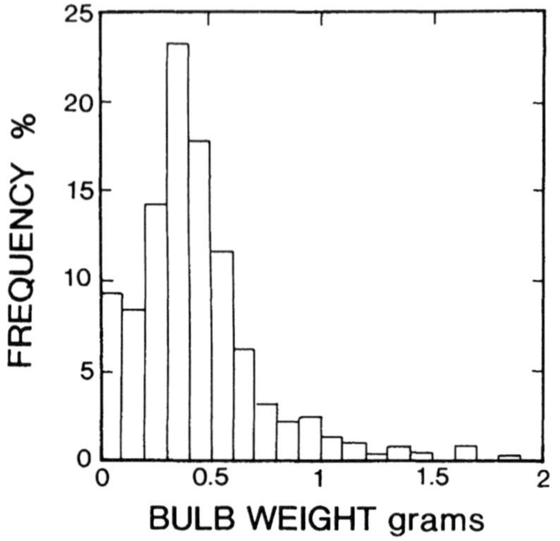

Fig. 7. Size distribution of *Lilium longiflorum* bulbs propagated in 2 liter bioreactors n, 634 (from 250 bulbscales) (Takayama and Fukano, in preparation).

or through bioreactor production. For these, it may be critical to obtain large bulblets *in vitro* in order to produce a blooming-size bulb in a reasonable length of time. Therefore, the response of the *Lilium* cv. "Enchantment" to increased levels of sucrose was examined using 20 liter bioreactors. Bulbscales were first cultured in shake culture in a medium with 3% sucrose, 1× MS salts, 2 mg/liter kinetin, and 0.1 mg/l NAA. Bulblets formed in these cultures were transferred to 20 liter bioreactors containing the same medium and cultured for 8 weeks. An inoculum of 25 bulblets derived from these bioreactors were transferred to bioreactors containing $\frac{1}{2}$× MS salts, no hormones, and either 3 or 9% sucrose. Bulbs were harvested and sorted into four arbitrary classes based on bulb diameter. Class 1 bulbs were less than 1.0 cm in diameter, class 2 between 1.0 and 2.0 cm, class 3 between 2.0 and 3.0 cm, and class 4 greater than 3.0 cm.

The results from the harvest of five bioreactors from each of the sucrose concentrations are presented in Table I. After 8 weeks, 330 bulblets were harvested and classified from medium with 3% sucrose and 270 from medium with 9% sucrose. The size of the bulbs produced was greatest in medium with 9% sucrose, and the largest bulbs obtained were approximately 4 cm in diameter. However, these large bulbs were rare and the majority of the bulbs formed in 9% sucrose were of the class 2 variety (1.0–2.0 cm). There is clearly a shift in distribution toward the production of larger bulbs when sucrose concentration is increased from 3% to 9%. Cultures grown in 3% sucrose exhibited more scale leaf formation than cultures grown in 9% sucrose. This is consistent with previous observations (Takayama and Misawa, 1982a). Similar effects have been observed with most *Lilium* hybrids tested, including both Asiatic and Oriental types. However, for some cultivars, 6% sucrose was optimal.

In a separate experiment, bulblets of Red Carpet were produced in a similar manner on 9% sucrose, harvested, and sorted into two size classes: 0.5–1.5 cm diameter (small) or 2.0–3.0 cm diameter (large). Bulblets lying outside these ranges were not used in these experiments. Fifty bulblets were included in each size class, and the experiment was repeated once. These bulbs were planted in the soil in the greenhouse and treated in an attempt to induce flowering within 1 year of transplant to the soil. Survival of transplants in the first experiment was 100%, while in the second experiment survival was 94% for the small class and 100% for the large class. In the first experiment, none of the bulbs in the small class flowered during the first year, but 62% of the large bulbs flowered. The second experiment produced the same results for the small bulbs and 48% flowering for the large bulbs.

It is apparent that inducing larger bulbs to form *in vitro* can hasten the maturation of the bulbs and that this *in vitro* increase in size of the bulbs can be induced by using 6 or 9% sucrose. However, it has yet to be determined what the optimal conditions are for inducing consistent increases in bulb size and for consistent flowering. As can be seen in Table I, the range in size of bulbs produced in bioreactors is fairly high. Sorting of the bulblets after harvest can probably result in consistent responses in growth in the soil, but it would be more cost effective if this sorting were not necessary and bulb size were more uniform.

TABLE I

Size Distribution of Bulbs of *Lilium* cv. Enchantment Harvested from Bioreactors Containing 3% or 9% Sucrose[a]

Sucrose conc. (%)	Total no. bulbs	No. bulbs harvested per category (% of total)			
		$d < 1.0$[b]	$1.0 < d < 2.0$	$2.0 < d < 3.0$	$d > 3.0$
3	330	172 (52%)	119 (36%)	39 (12%)	— (0%)
9	270	75 (28%)	116 (43%)	63 (23%)	16 (6%)

[a] Initial inoculum was 125 bulbs for each concentration of sucrose. Cultures were harvested, counted, and graded 2 months after inoculation.

[b] *d*, diameter of bulbs in centimeters.

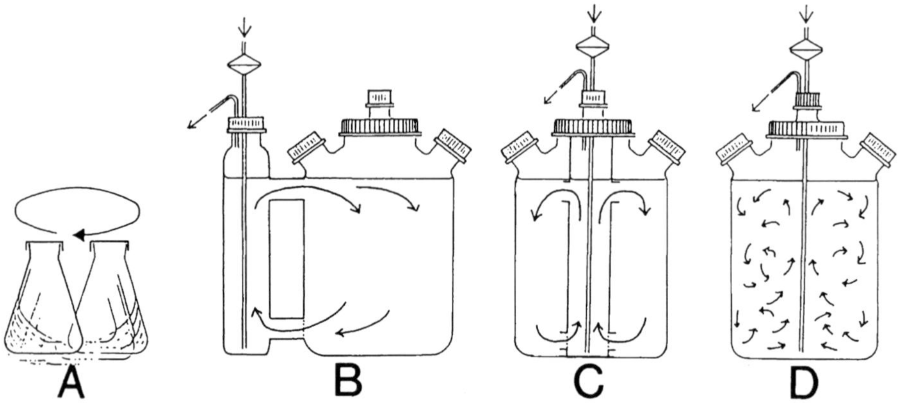

Fig. 8. Various types of bioreactors used for *Lilium* propagation. (A) Shake flask; (B) and (C) air lift bioreactor; (D) simple aeration bioreactor.

C. Bioreactor Construction

Various types of bioreactors (aeration–agitation, air driven, rotating drum, spin filter, gaseous phase, or introduced light) have been used for plant tissue cultures (for review see Takayama, 1990, 1991). Among these, we have used two types of simple air-driven bioreactors (Fig. 8). Air-driven bioreactors have an air sparger at the bottom of the vessel. The sparger serves both to move the propagules about in the medium and to aerate the cultures. This type of reactor has proved to be excellent for both plant cell and plant organ cultures. In *Lilium* microbulb propagation, this type of bioreactor is practical because of the low price of construction and easy handling. Vessels can take many shapes and be of virtually any dimension, provided that aeration is adequate. However, we have found that a cylindrical bioreactor is typically the easiest to use, as aeration in corners is not a problem and construction is easy. Obviously, scale-up to very large vessels presents problems for aeration, movement of propagules, sterilization of the vessel, and simple handling of the vessel (moving it from place to place). The problems are also compounded by the level of disaster associated with losses due to contamination. Large vessels with large inocula result in large losses due to the introduction of only one contaminant. While a 5% contamination rate of propagules may be considered acceptable in a traditional micropropagation scheme where most propagules are kept separate, if 5% of propagules are contaminated and 20 propagules are used to inoculate cultures, the chances of success are extremely limited. For the above reasons, the most feasible size of bioreactor to use at this point is between 2 and 20 liters. Advances in automation of the processes should help to eliminate problems with contamination and make it easier to manipulate larger containers.

IV. MICROPROPAGATION FROM BULBSCALES OF MICROBULBS IN NONSTERILE CONDITIONS

Attempts have been made to simplify the bioreactor process while maintaining high levels of multiplication. Trials have been made to regenerate bulbscales in water in an open-vessel aerated by compressed air through an air sparger as for conventional closed-containers. Small microbulbs have been produced using this method. The percentage of micropropagated bulbscales producing microbulbs varied from experiment to experiment but ranged between 70 and 100%. The size of these microbulbs was dependent on the size of the bulbscale used to initiate the cultures and ranged from 2 to 5 mm in diameter. These microbulbs have small-scale leaves about 10–20 mm in length. Microbulbs obtained by this process can be planted by simply spreading them onto a layer of soil, burying with a thin layer of soil, and covering with a nylon sheet to prevent desiccation. After 2–4 weeks, microbulbs produce scale leaves and bulb growth begins.

Bulbscales from *in vitro* cultures have also been used to produce microbulbs directly in the soil (Fig. 9). The process is something of a hybrid between traditional bulbscale cutting methods and tissue culture techniques. Microbulbs were regenerated from 100% of *in vitro*–produced bulbscales. However, germination of these bulblets was only 30–50%. It would seem that some additional nutritional or hormonal treatments will be required to make this a viable alternative to production of bulblets *in vitro*.

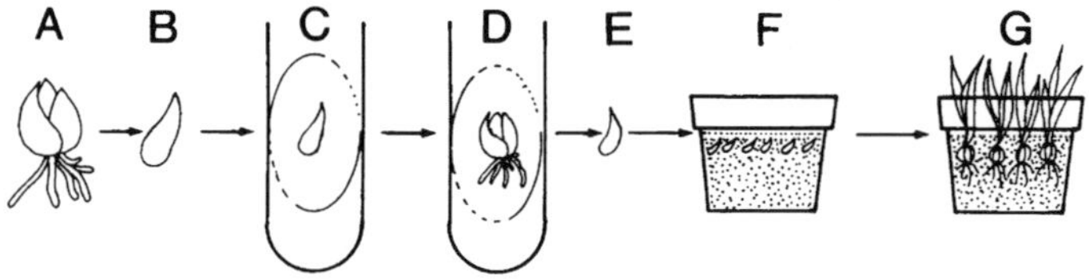

Fig. 9. Steps for propagation by nonsterile bulbscale cutting from bulbscales propagated *in vitro* (Takayama and Fukano, in preparation).

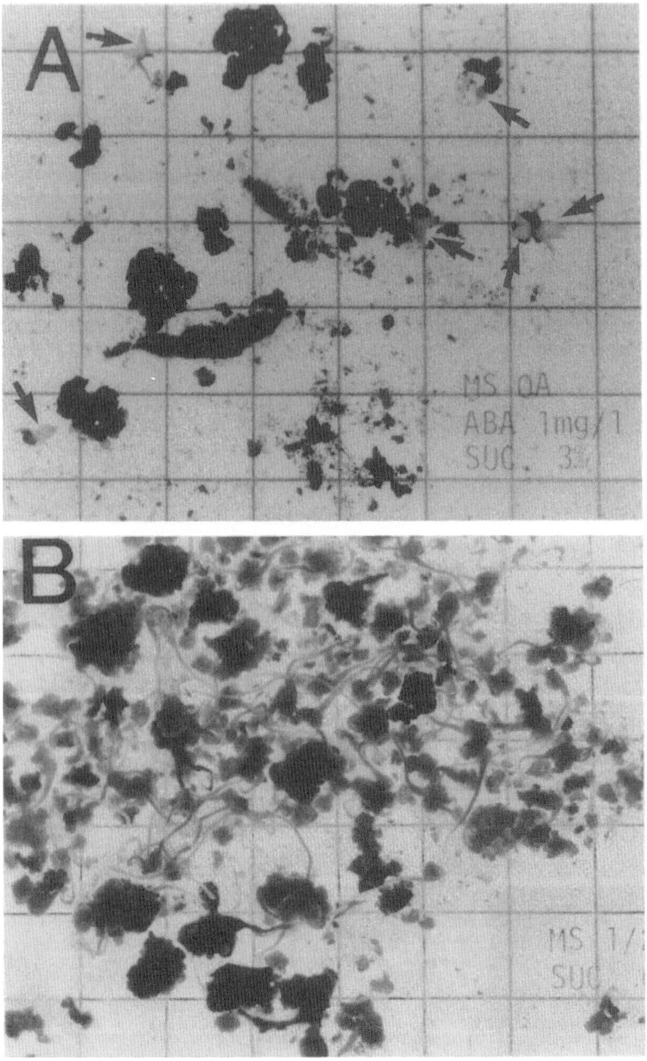

Fig. 10. Regeneration of microbulbs from suspension cultured cells of *Lilium longiflorum*. (Takayama and Fukano, in preparation) (A) regeneration of microbulbs in modified MS medium (NH_4NO_3, KNO_3, and $CaCl_2{\cdot}2H_2O$ deleted), MS supplemented with 1 mg/liter ABA and 30 g/liter sucrose. Cell growth was inhibited, but, microbulb production was stimulated. (B) Regeneration of roots in ½ strength MS medium supplemented with 60 g/liter sucrose. Cell and root growth was stimulated but no microbulbs formed.

V. MICROPROPAGATION FROM SUSPENSION-CULTURED CELLS

Regeneration of microbulbs from callus has been reported previously (Simmonds and Cumming, 1976; Stimart *et al.*, 1980). In many plant species, addition of auxin to the medium stimulates callus formation, but in *Lilium* species, auxin is not a fundamental prerequisite for callus induction. Addition of NAA or 2,4-dichlorophenoxyacetic acid (2,4-D) at concentrations of 1, 3, 10, or 30 mg/liter did not induce callus formation in *L. longiflorum*. However, callus induction was stimulated by the following conditions: (1) increased levels of ammonium nitrate in MS medium, (2) oxygen-deficient conditions, (3) high osmotic potential, and (4) increased temperatures. Once callus cultures were established on agar medium, cell cultures were easily established. Regeneration of microbulbs from suspension cultured cells was quite difficult, however. A number of experiments involving combinations of different salt concentrations and formulations, different sucrose concentrations, and addition of various levels of ABA revealed that deletion of ammonium nitrate, potassium nitrate, and calcium chloride from MS salts and addition of 3% sucrose and 1 mg/liter ABA stimulated the regeneration of microbulbs (Fig. 10A) (S. Takayama and M. Fukano, unpublished results). Other combinations such as $\frac{1}{2}\times$ MS salts with 6% sucrose without ABA stimulated rhizogenesis (Fig. 10B).

VI. ROBOTICS IN *LILIUM* MICROPROPAGATION

A. Introduction

A fully automated system for micropropagation would be useful because labor costs occupy the majority of total production costs of *in vitro*-propagated plants. Murase and Miwa (1989) developed a robot controlled by a microcomputer named "Mericlone Robot" to be used for automated plant tissue culture. This robot has been used for aseptic transfer of plant tissues *in vitro*. It is operated by a TiNi shape memory alloy actuator and a microcomputer. A more detailed outline of the system follows.

B. General Design Concepts

In conventional micropropagation for lilies on an agar medium, the subculture usually proceeds along the following lines: (1) removing roots that have formed, (2) separating the bulbscales of newly formed microbulbs, and (3) transferring those bulbscales into a culture vessel with fresh media. The initial plan for the system called for an attempt to address these three major steps. Problems that were anticipated and encountered include: (1) variability in the size and shape of *in vitro*–produced microbulbs could effect the efficiency of the automated operation of the system, and (2) roots would be of a variable size and create varying degrees of difficulty in the transfer process. Both of these obstacles were overcome with the design of the system (Fig. 11) and are discussed below.

C. Mechanical Construction, Mechanism, and Operation

When large numbers of microbulbs were removed from a vessel(s), the mass of tissues was generally a tangled mixture of roots and shoots that was difficult to separate into individual bulblets. Entangled bulbs were

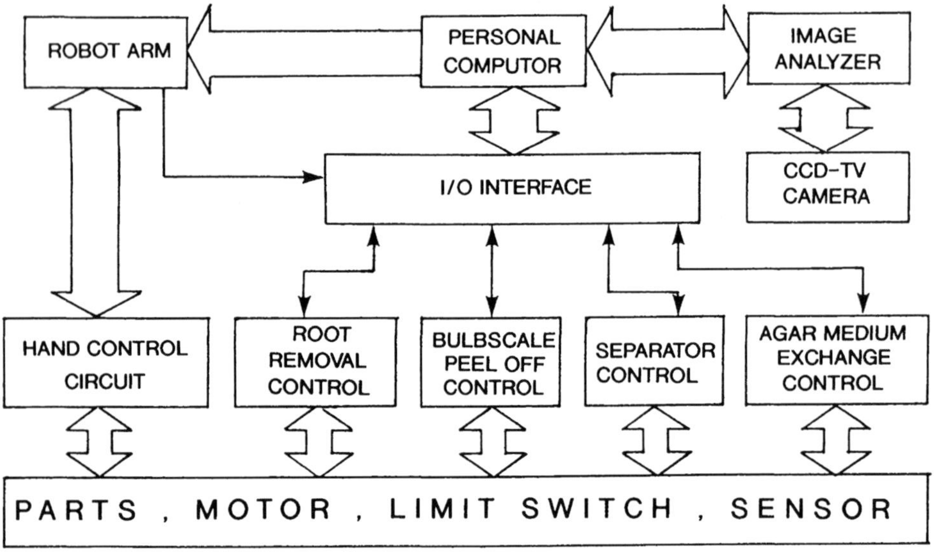

Fig. 11. Process flow diagram of the mericlone robot developed for automated *in vitro* transfer of *Lilium* bulbscales (Murase *et al.*, 1989).

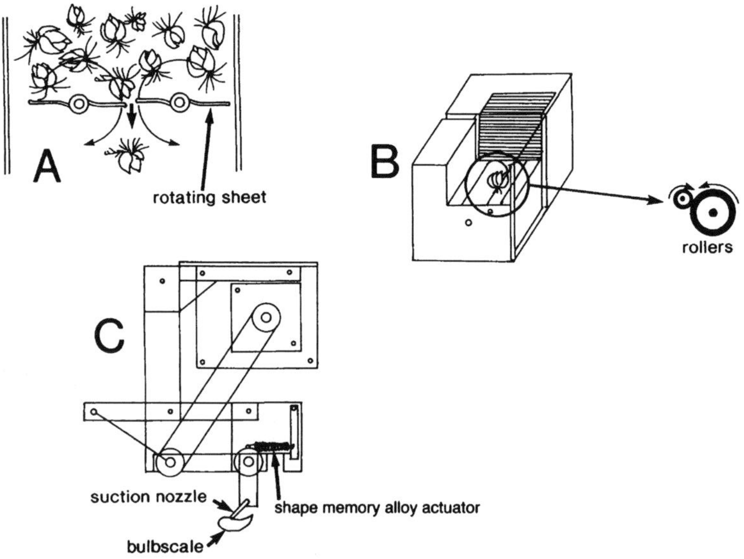

Fig. 12. Mechanical diagrams of the mericlone robot used for automated *Lilium* propagation. (A) Bulb supplier; (B) Root remover; (C) robot hand (Murase *et al.*, 1989).

separated in two steps, first by dropping little by little from a rotating sheet (Fig. 12a) onto a vibrating tray, and then by separating with this vibrating tray. At this point individual bulblets could be transferred to the mechanical apparatus designed to remove the roots.

Removal of the roots was accomplished by rolling the bulblets on rollers which engulfed and removed the roots but kept the body of the bulblet intact (Fig. 12b, left). Various combinations of parallel rollers were tested, and a combination of 11 and 4 mm diameter rollers proved to be optimal for this process (Fig. 12b, right).

Microbulbs were then transferred to a disk rotating at a high speed (10 cm diameter, 5000 rpm). Microbulbs were forced against the side walls surrounding the disk, and bulbscales were peeled off without severe damage within 5 sec. Bulbscales were then supplied one by one to a robot hand (Fig. 12c), operated with the aid of an image analyzer. The robot hand was used to transfer bulbscales to agar medium automatically. In this process, the proximal end of the bulbscale was detected within 2 sec by image analyzer calculation from two dimensional image data. The bulbscales were sucked up in a suction nozzle and gently grasped by a springy shape memory alloy actuator and transferred onto the agar medium without damage.

VII. AUTOMATED PROCESS FOR TRANSPLANTATION OF *LILIUM* MICROBULBS INTO SOIL

Automated processes for transplanting *Lilium* microbulbs into the soil have not been established, but some simple mechanical processes should be applicable to this procedure. Essentially, the separation process described for automatic separation of bulblets for subculturing could be taken advantage of for a similar process. Before the root removal stage, bulblets could be dropped or placed onto a layer of soil. Another layer of soil could then be dispensed to cover the propagules and the transplantation process would be complete. A similar device as described above for image analysis and subsequent orientation of the bulblets may be necessary to obtain optimal growth, but this should also be something of an extension of the processes developed above.

VIII. IDEALIZED AUTOMATED MICROPROPAGATION PROCESSES FOR *LILIUM* IN THE FUTURE—PROBLEMS AND PERSPECTIVES

Microbulbs of *Lilium* species formed *in vitro* in rapid multiplication processes typically have a maximum size of about 1 cm in diameter. The size and morphology of these bulbs can be quite variable, and this characteristic is one of the main barriers to the establishment of a fully automated propagation process. The Mericlone Robot, consisting of an actuator, a microcomputer, and an image analyzer, has been developed to help overcome some of the problems associated with the variability encountered and make it possible to develop an automated system. However, at present the Mericlone Robot is extraordinarily expensive. Additionally, its actual operation speeds are too slow and problematic for retention of long-term sterile conditions for an economically viable industrial process. Although the development of robot technology has been rapid and will eventually become an important tool in commercial micropropagation, at this time a simpler system is required.

In this context, we believe that the information presented above is an important step in the development of an efficient, cost-effective, automated system for micropropagation. This step is the development of bioreactor technology. It is clear that this technology will have an important impact in the near future in providing cost reductions in the micropropagation process.

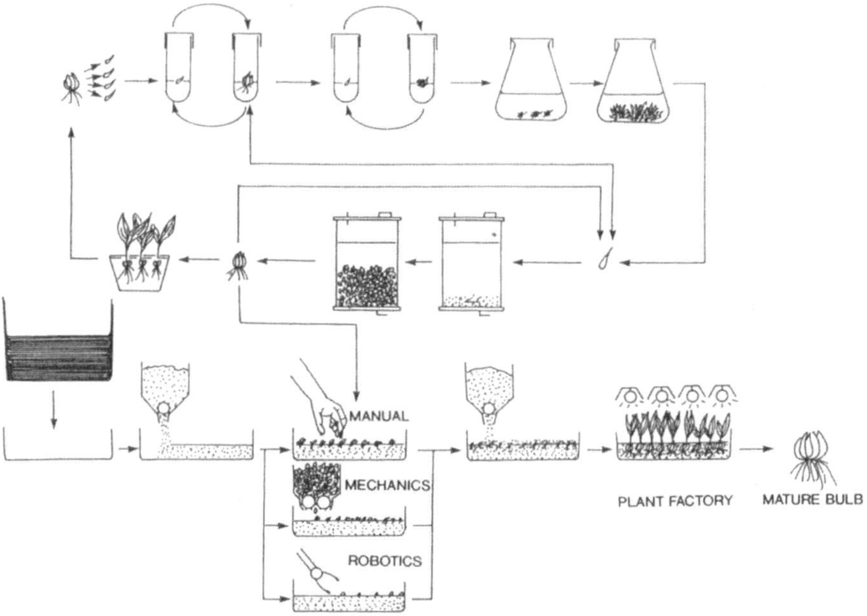

Fig. 13. Schematic of an idealized, complete system for automated propagation of *Lilium*.

We present in Fig. 13 an idealized system for automated propagation of *Lilium* species and hybrids. This idealized view is close to realization and suffers mostly from the cost of the development and operation of the equipment involved. However, with continued efforts in these areas, we feel there is little doubt that cost reductions will be achieved in these areas and the equipment will be steadily improved to yield a viable, automated system or group of systems applicable to a number of plant species.

REFERENCES

Allen, T. C. (1974). Control of viruses in Lilies. "Lilies 1974 and other Liliaceae." pp. 3–10. Robert Dinwiddle & Co. Ltd., Dumfries, Scotland.

Allen, T. C., and Fernald, K. (1972). Elimination of viruses in hybrid lilies. *Lily Yearbook, N. Amer. Lily Soc.* **25,** 53–55.

Allen, T. C., and Linderman, R. G. (1976). Rate of infection of virus-free lilies with lily symptomless virus in the field. *Acta Hortic.* **59,** 37–38.

Anderson, W. C. (1977). Rapid propagation of *Lilium* cv. Red Carpet. *In Vitro.* **13,** 145.

Bennici, A. (1979). Cytological chimeras in plants regenerated from *Lilium longiflorum* tissues grown *in vitro*. *Z. Pflanzenzüchtg.* **82,** 349–353.

D'Amato, F. (1977). Cytogenetics of differentiation in tissue and cell cultures. *In* "Applied and Fundamental Aspects of Plant Cells, Tissue and Organ Culture" (J. Reinert and Y. P. S. Bajaj, Eds.), 343–356. Springer-Verlag, Berlin.

Hansch, C., Muir, T. M., Fujita, T., Maloney, P. P., Geiger, F., and Streich, M. (1963). The correlation of biological activity of plant growth regulators and chloromycetin derivatives with Hamet constants and partition coefficients. *J. Amer. Chem. Soc.* **85,** 2817–2824.

Hansch, J., and Fujita, T. (1964). ρ-σ-π analysis. A method for the correlation of biological activity and chemical structure. *J. Amer. Soc. Hort. Sci.* **86,** 1616–1626.

Hartmann, H. T., and Kester, D. E. (1975). "Plant propagation: Principles and practices." 3rd Ed. Prentice Hall, Inc., Englewood Cliffs, New Jersey.

Liu, L., and Burger, D. W. (1986). *In vitro* propagation of Easter lily from pedicels. *HortSci.* **21,** 1437–1438.

Murase, M., and Miwa, Y. (1989a). Studies on automechanization of culture systems for microbulb production of lilies. *Abst. Jap. Soc. Mechanical Engineers,* No. 890–38, pp. 38–39.

Murase, M., Kodama, K., Sugiura, M., Ito, A., Miwa, Y., and Nakagawa, M. (1989b). Studies on plant tissue cultures: Development of an equipment for removal and transplantation of bulbscales from microbulbs of lilies. *J. Japn. Soc. Precision Engineer.* **H-51,** pp. 627–628.

Murashige, T., and Skoog, F. (1962). A revised medium for rapid growth and bioassays with tobacco tissue cultures. *Physiol. Plant.* **15,** 473–497.

Mowat, W. P., and Stefanac, Z. (1974). Aphid-transmitted viruses from lilies in Britain. *Ann. Appl. Biol.* **76,** 281–287.

Niimi, Y. (1984). Bulblet-productivity of explants from scales, leaves, stems and tepals of *Lilium rubellum* Baker. *Sci. Hort.* **22,** 391–394.

Niimi, Y., and Onozawa, T. (1977). Vegetative propagation of *L. rubellum* Baker; comparison of bulblet formation between scaling and segment cultured *in vitro. Abstracts* (Fall Meeting) *Jpn. Soc. Hortic. Sci.* pp. 348–349 (in Japanese).

Novak, F. J., and Petru, E. (1981). Tissue culture propagation of *Lilium* hybrids. *Scientia Hortic.* **14,** 191–199.

Riviere, Mme. S. (1968). Culture *in vitro* du meristeme apical de la tige annuelle du *Lilium candidum,* Liliacees L. *C.R. Acad. Sci., Ser. D.* **267,** 1439–1441.

Robb, S. M. (1957). The culture of excised tissue from bulbscale of *Lilium speciosum* Thunb. *J. Exp. Bot.* **8,** 348–362.

Sharp, W. R., Raskin, R. S., and Sommer, H. E. (1971). Haploidy in *Lilium. Phytomorphol.* **21,** 334–337.

Sheridan, W. F. (1968). Tissue culture of monocots. *Lilium. Planta* **82,** 189–192.

Simmonds, J. A., and Cumming, B. G. (1976). Propagation of *Lilium* hybrids. II. Production of plantlets from bulb-scale callus cultures for increased propagation rates. *Scientia Hortic.* **5,** 161–170.

Skoog, F., and Miller, C. O. (1957). Chemical regulation of growth and organ formation in plant tissues cultured *in vitro. Symp. Soc. Exp. Biol.* **11,** 118–130.

Stimart, D. P., and Ascher, P. D. (1978). Tissue culture of bulbscale sections for asexual propagation of *Lilium longiflorum* Thunb. *J. Am. Soc. Hortic. Sci.* **103,** 182–184.

Stimart, D. P., Ascher, P. D., and Zagorski, J. S. (1980). Plants from callus of the interspecific hybrid *Lilium* "Black Beauty". *HortSci.* **15,** 313–315.

Takayama, S. (1991). Bioreactors for plant cell, tissue and organ cultures. In "Handbook of Fermentation Biotechnology" Noyes Publishing Co., New Jersey. (in press).

Takayama, S. (1991). Mass propagation of plants through shake- and bioreactor culture techniques. *In* "Biotechnology in Forestry and Agriculture" Vol. 15 (Y. P. S. Bajaj, ed.) (in press).

Takayama, S., and Misawa, M. (1979). Differentiation in *Lilium* bulb-scales grown *in vitro.* Effect of various cultural conditions. *Physiol. Plant.* **46,** 184–190.

Takayama, S., and Misawa, M. (1980). Differentiation in *Lilium* bulb-scales grown *in vitro*. Effects of activated charcoal, physiological age of bulbs and sucrose concentration on differentiation and scale leaf formation *in vitro*. *Physiol. Plant.* **48,** 121–125.

Takayama, S., and Misawa, M. (1981). Mass propagation of *Begonia* x *hiemalis* plantlets by shake culture. *Plant Cell Physiol.* **122,** 461–467.

Takayama, S., and Misawa, M. (1982a). Regulation of organ formation by cytokinin and auxin in *Lilium* bulbscales grown *in vitro*. *Plant Cell Physiol.* **23,** 67–74.

Takayama, S., and Misawa, M. (1982b). A scheme for mass propagation of *Lilium in vitro*. *Scientia Hortic.* **18,** 353–362.

Takayama, S., and Misawa, M. (1982c). Mass propagation of ornamental plants through tissue culture. *Proc. 5th Int. Cong. Plant Tissue Cell Cult.* pp. 681–682.

Takayama, S., and Misawa, M. (1983). The mass propagation of *Lilium in vitro* by stimulation of multiple adventitious bulbscale formation and by shake culture. *Can. J. Bot.* **61,** 224–228.

Takayama, S., Misawa, M., Takashige, Y., Tsumori, H., and Ohkawa, K. (1982). Cultivation of *in vitro* propagated *Lilium* bulbs in soil. *J. Amer. Soc. Hort. Sci.* **107,** 830–834.

Takayama, S., Fukano, M., and Amo, T. (1986). Studies on micropropagation of Lilies. Part 9. Stimulative effect of ABA on bulblet formation. Abst. (Spring Meeting) *Jap. Soc. Hort. Sci.* 336–337.

van Aartrijk, J., and Blom-Barnhoorn, G. J. (1981). Growth regulator requirements for adventitious regeneration from *Lilium* bulbscale tissue *in vitro*, in relation to duration of bulb storage and cultivar. *Scientia Hortic.* **14,** 261–268.

CHAPTER 7

Problems and Perspectives for the Automated Propagation of Fruit Trees

Alessio Martinelli

Consorzio Italiano Vivaisti
44020 S. Giuseppe di Comacchio (Ferrara)
Italy

I. INTRODUCTION

Fruit species, particularly the trees, are still produced in small amounts in commercial micropropagation when compared with the production of herbaceous ornamentals. This is partly due to the fact that they are more difficult to produce and also to the relative low price of the small micropropagated plants on the market. Conventional propagation is usually cheaper than tissue culture, and conventionally propagated plants often represent a better product for the grower. In particular, this is true in regard to the "juvenility" of the micropropagated plants and to the need for an expensive acclimation and hardening off phase, together with the need for long-term field evaluation of plants produced in tissue culture.

Until now fruit tree micropropagation has been principally used for the rapid propagation of virus-indexed material in order to produce mother plants for further propagation, as in the case of strawberry (Boxus, 1989), and apple and pear clonal rootstocks, or to produce rootstocks that are difficult to propagate by conventional means, as is the case with peach rootstocks like the French and American peach–almond hybrids (Zimmerman, 1986; Hammerschlag, 1986; Fiorino and Loreti, 1987).

Finally, micropropagation plays an important role in the rapid introduction of new genotypes and in the international exchange of material.

The key factors that can improve the size of the fruit micropropagation industry are cost reduction and amplification of the market. Cost reduction is linked, particularly for fruit trees, to the solution of some basic problems such as low rooting, leaching of toxic substances, vitrification, low multiplication rate, difficult acclimation, and of some technological problems such as the lack of repeatability and standardization, the high labor input, and the difficulties in scaling up.

In order to increase the market for micropropagated fruit plants, conditions must be developed for: (1) production of the final plant for the growers, (2) production of self-rooted scion cultivars, (3) production of plants with "added" value or characteristics, and (4) development of new cultivars and of new cultural techniques compatible with micropropagation.

The usefulness of automation in the improvement of fruit plant micropropagation will depend on the development of new processes leading to the satisfaction of these requirements.

II. TECHNOLOGICAL ASPECTS RELATED TO AUTOMATION

A. Brief Description of the Actual Processes

Fruit tree micropropagation is mostly based on enhanced axillary branching through the development of shoots from lateral buds. For some species, like olive and grapes, the method of nodal production is used (Rugini, 1986). Both systems are characterized by high genetic stability but a low multiplication rate. Genetic stability is crucial in fruit tree mass propagation, and there is no level of acceptable variation.

Transfer intervals range between 2 and 4 weeks and multiplication rate, calculated on a base of 10–12 cycles, can vary between 1.8 and 3.5 to 1 per transfer for tree species. Higher rates are obtained with shrubs and herbaceous fruit plants, like strawberry, raspberry, blackberry, and blueberry. An elongation phase between shoot proliferation and rooting is a common practice in order to harvest more and better shoots for rooting. Shoot elongation is obtained by placing shoots in a medium with reduced cytokinin or by adding a liquid medium in the jar at the end of the multiplication period (Maene and Debergh, 1985).

Rooting is principally done *in vitro*. Several attempts have been made to root shoots in the greenhouse, but results have been inconsistent, depending on the season and on the physiological state of the shoots coming from the laboratory. *In vivo* rooting applied commercially has been reported for walnut (Driver, 1985) and for pistachio (Martinelli, 1988). A simplified method of rooting apple cultivars *ex vitro* in plugs after an induction phase *in vitro* without agar has been reported by Zimmerman and Fordham (1985). In the commercial laboratories the *in vitro* phases are generally carried out in containers of 0.5 to 1 liter capacity. Most of the laboratories use glass jars. Several plastic containers have been tested, but results are good and consistent only for some genotypes and only in some steps. In the current practice, glass jars are filled with the medium and then autoclaved. Changes in the dispensing practice can be expected by the development of systems using presterilized disposable plastic containers and by the development of methods for autotrophic culture. Currently jars inoculated with the cultures must be sealed with a transparent film to avoid evaporation and contamination. This affects the growth of the plants by limiting gas exchange and by increasing the relative humidity in the vessel.

The most common practice for acclimatization is to transplant rooted plantlets in a mixture of peat and perlite, in plastic trays, and then to put the trays under a polyethylene tent at very high relative humidity (RH). After 10–15 days the RH is progressively reduced by opening the tunnel for increasing lengths of time. All these operations must be done by hand and require careful attention, accounting for the particularly high labor costs in fruit tree micropropagation.

B. Cost Analysis

Sluis and Walker (1985) reported a general analysis of costs in micropropagation resulting in breakdown of expenses as follows: 40% labor,

10% materials, 20% overhead, and 30% as sales, general, and administrative expenses. Other authors reported an incidence of labor ranging between 60 and 85% of the total cost (Chu, 1986; Donnan, 1986; see also Chapters 1 and 2, this volume).

As for fruit trees, the amount of labor required for transfer in the laminar flow cabinet is the most costly factor. There is also a major influence of the acclimation phase on the total cost of the finished plants. This is due to the lack of standardization of the transfer cycles and to the more difficult rooting of fruit trees compared with most of the herbaceous ornamentals. A finished tissue culture plantlet after the acclimation period can be sold for 40 cents for the rootstocks and 80 cents for some self-rooted cultivars.

C. Repeatability and Standardization

One typical feature of commercial fruit tree micropropagation is the lack of uniformity and repeatability of the different culture cycles. Part of this is caused by such problems as "vitrification" (Pàques and Boxus, 1987), bacterial contamination, phenolics exudation, and aging. But other factors, linked to the physiology and to the "ecology" of the cultures, are also involved. These factors are not known in detail yet, but they limit the development of the technology and its adaptability to automation.

In practice a manager of a fruit tree micropropagation laboratory is continually faced with the need to adjust the technology for production at almost every multiplication cycle by playing on the different factors and their connections.

D. Interconnections of the Different Factors

Several technological factors combine to determine the success of a commercial micropropagation process, and all of them influence each other, forming an interconnected network of parameters with which to work in order to get satisfactory production in terms of quality and cost. Some of these factors are very critical and merit further analysis.

1. Sterility

Micropropagation, as all tissue culture techniques, is based on the concept of sterility: cultures are grown in sterile sealed vessels, with sterilized media, and operations take place in sterile laminar flow cabinets. The implications of sterility on the process are several. First, there is the need for special facilities and equipment, with high investment. Then all the operations require much attention and are more complicated, both for human operators and for possible robots. Another very important implication is the limitation of the container, which must not be too large and must be sealed. Therefore, each jar is a small microenvironment that can differ from the others. These problems become more relevant in the commercial environment, where a large number of cultures must be handled daily.

2. Culture Vessels

While at the beginning of the micropropagation industry more emphasis was given to the culture media, the influence of the environment and, particularly, of the container is becoming more and more recognized. Owing to the need for sterility, the container must be small and sealed, with several implications:

1. Media preparation and washing operations cost much more using smaller jars. More support workers are needed than laminar flow hood worker.
2. There is a positive correlation between the size of the jar, the number of cultures per jar, and the number of cultures handled per operator day. Using larger vessels would make it possible to increase the number of transfers per operator day.
3. Mechanical access to the shoots being separated is difficult in the deep vessels commonly used.
4. The quality of the shoots arising in culture is highly influenced by the container. The amount of CO_2, ethylene, and ethanol in the jars during the growth period (Thomas and Murashige, 1979) condition the kind of development of the shoots. Relative humidity and its gradient has also been demonstrated to influence culture growth, particularly in reference to vitrification and to shoot tip necrosis (Vanderschaege and Debergh, 1987). All these parameters vary from jar to jar and are difficult to keep controlled and uniform.

3. Number of Transfers Per Operator Day

The number of transfers or explants that an operator can make in one day is clearly of great impact on the production cost. As mentioned before, the size of the container and the need for sterility limit the possibility of increasing this figure, which for fruit plants ranges between 1600 and 3000 explants per day, depending on the species and the quality of the cultures.

In the case of fruits, not only the number but also the kind of transfer is critical. In fact, the cut to be made in the cluster of shoots varies during the production process. Managers of production laboratories have learned that when the cultures are aging they need to be cut in smaller explants, or that sometimes only a little "pruning" is needed in order to get good shoots growing. This empirical knowledge is still very important and makes the development of automated transfer operations more difficult.

III. APPROACHES TO AUTOMATION OF FRUIT TREE MICROPROPAGATION

The problem of automation in the fruit micropropagation industry is focused on two questions: cost reduction and development of "new" products, where "new" means plants that for their characteristics and for their cost can go directly to the final user, considerably increasing the size of the potential market. This is particularly true for the strawberry, for the apple rootstocks, and for scion cultivars on their own roots.

A. Automation of the Actual Procedures

1. Media Preparation

Media preparation and dispensing is an area in which automation can be introduced in the actual processes. Devices used in microbiology laboratories can be adapted, and some commercial laboratories have developed their own equipment.

A special device for micropropagation has been introduced by a commercial firm in Holland. It is a system in which moving sterile plastic boxes are automatically filled in a laminar flow cabinet. The system is adapted for the possibility of having a robot place the explants in the jars before they are capped. The dispensing device costs about $110,000.

2. Transfer of Explants under Laminar Flow Hood

Since the beginning of the micropropagation industry, the idea of having robots doing the transfer work in the laminar flow cabinets was present in the hopes and in the perspectives of laboratory managers. De Bry (1986) suggested and described the technical possibility of designing robots usable in plant tissue culture. The robots consist of a vision system to detect the position and direction of the shoots and the site to cut (see also Chapter 12, this volume). The information is elaborated by the software that drives an end effector to the shoot or to the culture to be operated. The end effector can be a multi-joint arm that takes the shoot, cuts it in the right position, and transfers the cut portion in the new vessel containing the fresh medium. The software must give flexibility to the system so that it can be used with different crops and in different situations. The difficult access to the cultures in the containers actually used and the miniaturization and delicacy of the shoots need a very fine and accurate mechanical arm.

Some prototypes have been designed and developed (Deleplanque *et al.*, 1985; Rowe *et al.*, 1987; Fujita, 1989), but up to now they have not been used for fruit trees. One of the problems to be solved is the need for standardization of the transfer explants. The best solution would be to use single node cuttings, but trials with apple, pears, and peach rootstocks have shown that, with the current technology, results are inconsistent. Single node cuttings need more time to develop new shoots, their growth is uneven, and these features increase with the number of cycles (A. Martinelli, unpublished observation).

Another problem in the use of robots in fruit tree propagation is the high investment needed, which requires production of very large number of plants to recover the cost of the equipment.

B. Development of New Processes Adaptable to Automation

1. Use of Liquid Medium

One of the key steps in developing new automated processes is the possibility of using liquid medium and exchanging the medium during culture in order to reduce the frequency of transfers. The possibility of growing plantlets in liquid medium was discussed by Hussey (1986). Liquid medium additions were found to be very beneficial at the end of the multiplication phase to improve elongation and the subsequent rooting *in vivo* (Maene and Debergh, 1985). Experiments with fruit trees have shown the same beneficial effects (Viseur, 1987).

Tisserat and Vandercook (1985) introduced an automated plant culture system, controlled by a microcomputer, in which plant explants remain in the same container in liquid medium. Aliquots of the medium are periodically removed and substituted with fresh medium. Results presented on different crops showed that beyond the saving in transfers, there was a significantly higher increase in the fresh weight and in the volume of the tissues cultured through this system than in the same agar solidified medium. A similar commercial system was described by Farrel (1987).

A variant of this method is the use of mist-feeding. Weathers and Giles (1987) presented a new bioreactor, Mistifier, in which medium is provided to the cultures in the form of mist. They reported that in tests with *Cinchona, Ficus, Brassica, Lycopersicon,* and *Daucus* the growth was at least as good as in agar boxes, and regenerated plants were always higher in quality than those grown in agar. A bioreactor for the commercial propagation of a large number of plants is the "Vitromatic" System introduced by Levin *et al.* (1988). This system is based on the rapid multiplication of large and compact meristematic clusters in liquid medium in a bioreactor. A bioprocessor separates, sizes, and dispenses individualized meristems into culture vessels. In 8 hours the bioprocessor can produce sufficient material for 100,000 propagules. The system can be coupled with a microprocessor-controlled transplanting machine, which transfers plants to a greenhouse at the rate of 12,000 plants/hour (Levin and Vasil, 1989). This system could be suitable also for fruit trees, but there are no reports on its use on them yet.

Will these methods be suitable for fruit tree micropropagation? First of all it is necessary to determine the reaction of fruit species to culture in liquid medium. A liquid medium phase was found to be beneficial for the elongation of peach–almond hybrid shoots (Martinelli, 1985). Medium used was MS (Murashige and Skoog, 1962) concentrated 2 times with no hormones. Such a high concentration of salts was necessary to control vitrification. In further trials designed to test the feasibility of shoot proliferation in liquid medium, it was impossible to control vitrification after several subcultures. Vitrification could be controlled only by increasing the exchange of air in the jar (A. Martinelli, unpublished) by opening the vessel in the laminar flow hood twice weekly. In order to automate the process, however, it is necessary to design containers in which it will be possible to control the atmosphere without external interventions. Changes in the medium composition will also be necessary to sustain a good and lasting growth of the shoots.

2. Shoot Hedges

Aitken-Christie and Jones (1987) reported a novel method for *in vitro* shoot production of *Pinus radiata*. Shoots were cultured as hedges for a period of up to 18 months without manual subculturing. They were fed by liquid nutrient replenishment twice weekly. Shoots were harvested monthly and were reported to be more elongated and healthier than the conventionally treated material.

This method is suitable for the automation of woody plant micropropagation. In preliminary experiments designed to test the shoot hedge production with peaches, we found that good shoots were obtained only in the first 2–3 harvests. After that, only a few new shoots developed and fewer of them were good for rooting. The first harvested shoots were longer and with better leaf expansion than shoots obtained through the normal procedures. These results indicate a good potential of this system, but more experimentation is needed to define the right media and the right cultural conditions necessary to grow fruit species as *in vitro* hedges.

3. Autotrophic Culture

Autotrophic culture (Kozai *et al.*, 1988; Fujiwara *et al.*, 1988; see also Chapter 11, this volume) is a system in which cultures are grown in the absence of sugar. Growth is sustained with high light intensity and CO_2 enrichment. This method offers several advantages in view of automation and in view of the quality of the plantlets produced. The possibilities of using loosely capped larger vessels and of controlling environmental parameters in the container are particularly interesting. Manual operations are reduced, and plantlets obtained are more resistant to stress and acclimatize better in the greenhouse.

Sugar-free tissue culture has been reported for strawberry (Grout and Price, 1987). For fruit trees further investigation is needed. Important features to verify are: (1) the possibility of reducing vitrification risks, (2) the possibility of using liquid medium, (3) increase of uniformity, (4) improvement of rooting and acclimatization, and (5) improvement of standardization.

4. Somatic Embryogenesis

The possibility of obtaining large numbers of somatic embryos from a small amount of medium in a bioreactor has been indicated as a method of automated production of very cheap clonal "seeds" (Lutz *et al.*, 1985). Evidence that somatic embryos are genetically stable has been reported by Vasil (1988). The list of plants that can regenerate somatic embryos is continually increasing, and it includes several fruit species. Somatic embryogenesis has been recently reviewed by Ammirato (1989). For fruit species, somatic embryogenesis has been reported from explants taken from the true embryo (Hansen and Durzan, 1986), from the cotyledonary nodes (Perez *et al.*, 1986), from endosperm (Tulecke *et al.*, 1988), and from leaves of young seedlings (Liu *et al.*, 1983). But information on obtaining good production of somatic embryos in fruit plants from mature explants is still insufficient, and concerns about genetic stability have limited research efforts in commercial laboratories on the use of somatic embryogenesis for micropropagation of fruit trees.

IV. CONCLUSIONS AND PERSPECTIVES

Notwithstanding the increasing number of reports on the automation of micropropagation, very little is published concerning the experimentation and development of automation for fruit tree propagation. The new techniques developed for woody plants should be reliable also for most fruit trees, but careful experimentation is needed before such systems can be introduced for commercial production. It is particularly important to define whether these new technologies can lead to an increase in the standardization of the process and whether production of plants that acclimatize better and that are adapted to the final grower can be produced.

Good perspectives are opened by the development of autotrophic culture, since it is possible to obtain cheaper plants from the laboratory with very low acclimation requirements. Furthermore, the absence of sugar decreases the risks of contamination and allows the implementation of new production technologies more suitable for automation, like: (1) larger vessels with loosely fitting caps, (2) the possibility of operating in the growth chamber instead of under the laminar flow cabinet, (3) the use of liquid medium, and (4) the possibility of standardizing the gas content in the vessel. As a transfer technique, the shoot hedging looks very promising for fruit trees. Most of these systems have not been tested yet on a sufficiently large scale on fruit trees, so further work will be necessary before commercial application.

ACKNOWLEDGMENTS

I wish to express my thanks to Dr. R. H. Zimmerman, of the U.S. Department of Agriculture, Beltsville, Maryland, for reviewing the manuscript.

REFERENCES

Aitken-Christie, J., and Jones, C. (1987). Towards automation: Radiata Pine shoots hedges *in vitro*. *Plant Cell., Tissue Organ Cult.* **8,** 185–196.

Ammirato, P. V. (1989). Recent progress in somatic embryogenesis. *Intl. Ass. Plant Tissue Cult. Newsletter* **57,** 2–16.

Boxus, P. (1989). Review of *in vitro* strawberry mass production. *Acta Hort.* **265,** 309–320.

Chu, I. Y. E. (1986). The application of tissue culture to plant improvement and propagation in the ornamental horticulture industry. *In* "Tissue Culture as a Plant Production System for Horticultural Crops" (R. H. Zimmerman, R. J. Griesbach, F. A. Hammerschlag, and R. H. Lawson, eds.), pp. 15–33. Martinus Nijhoff, Dordrecht, The Netherlands.

De Bry, L. (1986). Robots in plant tissue culture: an insight. *Intl. Ass. Plant Tissue Cult. Newsletter* **49,** 2–22.

Deleplanque, H., Bonnet, P., and Postaire, J. G. (1985). An intelligent robotic system for in vitro plantlet production. *Proc. 5th Int. Conf. on Robot Vision and Sensory Controls,* 29–31 Oct. 1985, Amsterdam, pp. 305–314 . IFS Conferences Ltd.

Donnan, A. (1986). Determining and minimizing production costs. *In* "Tissue Culture as a Plant Production System for Horticultural Crops" (R. H. Zimmerman, R. J. Griesbach, F. A. Hammerschlag, and R. H. Lawson, eds.), pp. 167–173. Martinus Nijhoff, Dordrecht, The Netherlands.

Driver, J. A. (1985). *Abstract* Direct field rooting and acclimatization of tissue culture cuttings. *In Vitro* **21** (3, Pt. 2), **57A.**

Farrel, M. A. (1987). Liquid medium system for plant propagation. *Proc. Moet-Hennessy Conf. Electronics and Management of Living Plants,* Montecarlo, pp. 41.

Fiorino, P., and Loreti, F. (1987). Propagation of fruit trees by tissue culture in Italy. *HortScience* **22(3),** 353–358.

Fujita, N. (1989). Application of Robotics to Masspropagation System. (Abs.) *Tissue Culture Ass. Annual Meeting abstracts.* p. 22 A. N. 47. Orlando.

Fujiwara, K., Kozai, T., and Watanabe, I. (1988). Development of a photoautotrophic tissue culture system for shoots and/or plantlets at rooting and acclimatization stages. *Acta Hort.* **230,** 153–158.

Grout, B. W. W., and Price, F. (1987). The establishment of photosynthetic independence in strawberry cultures prior to transplanting. Symposium Florizel 87, Belgium.

Hammerschlag, F. A. (1986). Temperate fruits and nuts. *In* "Tissue Culture as a Plant Production System for Horticultural Crops" (R. H. Zimmerman, R. J. Griesbach, F. A. Hammerschlag, and R. H. Lawson, eds.), pp. 221–236. Martinus Nijhoff, Dordrecht, The Netherlands.

Hansen, K. C., and Durzan, D. J. (1986). Somatic embryogenesis and morphogenesis in *Prunus Cerasus. Abstract Proc. XXII Intl. Hort. Congress,* Davis, California.

Hussey, G. (1986). Problems and prospects of *in vitro* propagation of herbaceous plants. *In* "Plant Tissue Culture and Its Agricultural Applications" (L. A. Withers and P. G. Alderson, eds.), pp. 69–84. Butterworth, London.

Kozai, T., Koyama, Y., and Watanabe, I. (1988). Multiplication of potato plantlets *in vitro* with sugar free medium under high photosynthetic photon flux. *Acta Hort.* **230,** 121–127.

Levin, R., Gaba, V., Tal, B., Hirsch, S., DeNola, D., and Vasil, I. K. (1988). Automated Plant Tissue Culture for Mass Progagation. *Bio/Technology* **6,** 1035–1040.

Levin, R., and Vasil, I. K. (1989). Progress in reducing the cost of micropropagation. *Intl. Ass. Plant Tissue Cult. Newsletter* **59,** 2–12.

Liu, J. R., Sink, K. C., and Dennis, F. G., Jr. (1983). Adventive embryogenesis from leaf explants of apple seedlings. *HortSci.* **18,** 871–873.

Lutz, J. D., Wong, J. R., Rowe, J., Tricoli, D. M., and Lawrence, R. H. (1985). Somatic embryogenesis for mass cloning of crop plants. *In* "Tissue Culture in Forestry and Agriculture" (R. R. Henke, K. W. Hughes, M. J. Constantin, and A. Hollaender, eds.), pp. 105–116. Plenum Press, New York and London.

Maene, L., and Debergh, P. (1985). Liquid medium additions to established tissue cultures to improve elongation and rooting *in vivo*. *Plant Cell., Tissue Organ Cult.* **5,** 23–33.

Martinelli, A. (1988). Factors affecting *in vitro* propagation of the peach. almond hybrids Hansen 2168 and Hansen 536. *Acta Hort.* **173,** 237–244.

Martinelli, A. (1988). Use of in vitro techniques for selection and cloning of different *Pistacia* species. *Acta Hort.* **227,** 436–437.

Murashige, F., and Skoog, F. (1962). A revised medium for rapid growth and bioassays with tobacco tissue cultures. *Physiol. Plant.* **15,** 473–497.

Pàques, M., and Boxus, Ph. (1987). Vitrification, Review of the Literature. *Acta Hort.* **212,** 155–166.

Perez, C., Rodriguez, R., and Tames, R. S. (1986). Regulation of asexual embryogenesis in filbert cotyledonary nodes: morphological variability. *Plant Sci.* **45,** 59–64.

Rowe, W. J., Rowe, N. M., and Roepe, H. P. (1987). The Application of Robotics for an Automated Tissue Transfer System. *Proc. Moet-Hennessy Conf. Electronics and Management of Living Plants,* Montecarlo, pp. 41.

Rugini, E. (1986). 9. Olive (Olea europea L.). *In* "Biotechnology in Agriculture and Forestry 1, Trees I" (Y. P. S. Bajaj, ed.), pp. 253–267. Springer-Verlag, Berlin.

Sluis, C. J., and Walker, K. A. (1985). Commercialization of plant tissue culture. *Intl. Ass. Plant Tissue Cult. Newsletter* **47,** 2–12.

Thomas, D., and Murashige, T. (1979). Volatile emissions of plant tissue cultures. I, Identification of the major components. *In vitro* **15,** 654–662.

Tisserat, B., and Vandercook, C. E. (1985). Development of an automated plant culture system. *Plant Cell., Tissue Organ Cult.* **5,** 107–117.
Tulecke, W., McGranahan, G., and Ahmadi, H. (1988). Regeneration by somatic embryogenesis of triploid plants from endosperm of walnut, Juglans regia L. cv. Manregian. *Plant Cell Rep.* **7,** 301–304.
Vasil, I. K. (1988). Progress in the regeneration and genetic manipulation of cereal crops. *Bio/Technology* **6,** 397–402.
Vanderschaege, A., and Debergh, P. (1987). Technical aspects of the control of the relative humidity in tissue culture containers. Symposium Florizel 87, Belgium.
Viseur, J. (1987). Micropropagation of pear, *Pyrus communis,* in a Double-Phase Culture Medium. *Acta Hort.* **212,** 117–124.
Weathers, P., and Giles, K. (1987). A Novel Method of *In Vitro* Culture Adaptable to Many Varieties of Plants. *Proc. Moet-Hennessy Conf. Electronics and Management of Living Plants,* Montecarlo, pp. 45.
Zimmerman, R. H. (1986). Propagation of fruit, nut and vegetable crops. Overview. *In* "Tissue Culture as a Plant Production System for Horticultural Crops" (R. H. Zimmerman, R. J. Griesbach, F. A. Hammerschlag, and R. H. Lawson, eds.), pp. 183–200. Martinus Nijhoff, Dordrecht, The Netherlands.
Zimmerman, R. H., and Fordham, I. (1985). Simplified method for rooting apple cultivars *in vitro*. *J. Amer. Soc. Hort. Sci.* **110,** 34–38.

CHAPTER 8

Bioreactors and Image Analysis for Scale-Up and Plant Propagation

Dana Cazzulino
Henrik Pedersen
Chee-Kok Chin

Department of Chemical and Biochemical Engineering and Department of Horticulture
Rutgers, The State University of New Jersey
New Brunswick, New Jersey

I. INTRODUCTION

Bioreactor technology provides the potential for producing large numbers of plants cheaply and efficiently (Ammirato and Styer, 1985). One goal is to produce plants more cheaply and efficiently than tissue culture on agar solidified medium or through the use of shake flask cultures. This is made possible by advances in bioreactor operation that enable researchers to control environmental conditions within the reactor through on-line monitoring and provides a uniform environment for the coordinated development of embryogenic material. Furthermore, large-scale operation can reduce labor costs, which translates into economic advantages in commercial process. At the present time, however, no large-scale process for the propagation of plants using bioreactors exists. Fortunately, scale-up and large scale operation are techniques of a well-established fermentation industry, and completely new procedures do not have to be developed. However, the process engineering requirements for cultures fostering plant cell proliferation and development have some factors that are quite different from those typically seen with bacterial or fungal systems. One of the most obvious differences concerns the varying morphologies possible in plant cell systems. Single cells, cell aggregates, and organized plant structures from different developmental stages of embryogenesis represent a few examples of plant forms that may be observed in culture.

These particular cell structures are observed in the plant tissue culture technique that has the greatest potential for large-scale production in liquid culture: somatic embryogenesis. It combines many characteristics of a propagation system desired by plant breeders, namely, fast turnaround, genetic uniformity, and disease-free plants. However, in order to make somatic embryogenesis a viable system for plant propagation, many challenges remain to be addressed. The heterogeneous mixture of cells, cell clusters, and embryos ranging from globular to torpedo, as well as the deformed embryos and clusters of embryos that are observed in a culture, need to be quantified for rational evaluation of the conditions promoting "good" development in culture. In addition, viable "true to type" embryos that will grow into plantlets outside the liquid culture have to be identified by, perhaps, morphological or biochemical criteria. These embryos need to be physically separated from the rest of the biomass in the culture, and conditions for synchronizing development or at least improving the yield of embryos growing into plantlets need to be found.

Some of these challenges have been addressed. A few groups have used size and density separation techniques in order to isolate specific embryo stages, to synchronize development of the embryos, or to improve the "quality" of the embryos (Fujimura and Komamine, 1979; Giuliano *et al.*, 1983; Lutz *et al.*, 1985; Redenbaugh *et al.*, 1987). Unfortunately, either precision in isolating the embryos is low or labor is high. Others have developed ways for assessing the viability of somatic embryos, by biochemical markers (Stuart *et al.*, 1985b) or by conversion frequency (Redenbaugh *et al.*, 1986; Redenbaugh *et al.*, 1987). However, biochemical markers may not always be reliable (Stuart *et al.*, 1985a), and conversion frequency is a laborious and time-consuming task (Stuart *et al.*, 1985b).

Embryo viability is best assessed by detailed morphological comparison with zygotic embryos (Ammirato, 1987). Fortunately, as embryos develop in embryogenic cultures, they are recognizable by morphologically distinct features seen easily under a microscope. Being similar in morphology to zygotic embryos (Steward *et al.*, 1958), somatic embryos of dicotyledonous plants have the same three stages of embryo development that have traditionally served as a classification system. These stages are referred to as globular, heart, and torpedo embryos (Halperin and Wetherell, 1964). Globular embryos are recognized as spherical masses with a smoother surface layer than the irregular shaped undifferentiated cytoplasmically dense cell clusters. Heart embryos can be distinguished by signs of cotyledon formation. Finally, torpedo embryos are distinguished from hearts basically by size and increased elongation along the root–shoot axis. Deviations from the zygotic course of events are more likely to result in somatic embryos that do not germinate or fully develop into plants; however, this has not been quantitatively tested. This type of work normally requires careful microscopic examination of individual embryos and, as a result, has been prohibitive on a large scale.

Nevertheless, the distinct morphological patterns that are easily discernible in somatic plant embryos are indeed quantifiable and can serve to make the examination process easier and probably more objective. Work by Schiavone and Cooke (1985), in which the shapes of embryos were described by simple parameters such as axial length, diameter, and cotyledon length, was one of the first attempts to quantify the embryo developmental stages rather than relying on the usual qualitative descriptions. These authors pooled their data into classes that fit the qualitative descriptions but did not take full advantage of various statistical techniques to render the pooled information distinct. Furthermore, their procedure, requiring the isolation, slicing, and fixation of embryos is very tedious and labor intensive and, therefore, discourages any attempts at routine or overall quantitative assessments of embryos in a culture system. Also, material is sacrificed and could not, of course, be selected and grown further.

One practical solution to the problem of characterizing the morphology of somatic embryos in culture is through the use of available image analysis technology. An image analyzer, a powerful tool for rapid, objective and quantitative recognition of morphology, has the capability of making detailed morphological comparisons routine in an automated fashion. The data obtained can be used to assess the effects of environmental changes on culture development and to help define appropriate bioreactor operating strategies. It is also reasonable to expect that the direct coupling of sophisticated computer imaging equipment with bioreactor operation will afford a means to significantly enhance embryo production and conversion frequency.

II. IMAGE ANALYSIS

The usefulness of image analysis has been recognized in the biological sciences. In the plant sciences, for example, it is being applied to seed selection (Zayas *et al.*, 1986) and root length measurement (Barnett *et al.*, 1987). In plant tissue culture, it has found applications in the growth determination of suspension cultures (Smith and Spomer, 1987), callus (Aitken-Christie *et al.*, 1987), and regenerated whole plants (Smith *et al.*, 1989). It also has been used in the determination of embryogenic cells (Grand d'Esnon *et al.*, 1988) and of somatic embryos (Cazzulino, 1989; Cazzulino *et al.*, 1990a; Grand d'Esnon *et al.*, 1988). In cancer research, where it is being used to distinguish between cell types, it has been firmly established as a powerful tool (Brugal *et al.*, 1986), and commercial use is made of the technology. In all these examples, the image analysis tools are used as an automated system substituting for manual operations and control.

Since each situation has unique circumstances, the requirements that will provide a fully automated system vary. Image analyzers come in a number of configurations with differing capabilities. The capability of any particular system is, of course, the result of the associated hardware and software, and through refinements in the software the image analyzer can usually be made to automatically perform the analyses desired. This analysis requires identifying embryos in an embryogenic culture that contains a heterogeneous population of embryos and aberrant structures and determining the developmental stage of the embryo selected.

A. Hardware

The basic hardware components include a video camera, TV monitor, and a computer with an imaging board. For microscopic observation, the video camera is attached to a microscope, and the imaging board provides the enabling technology for analyzing the video images. The other hardware components affect three key variables: speed of analysis, accuracy of resolving an observation, and storage of raw data. These variables mark the constraints of the hardware and are important in determining which system is appropriate for the job required. The image is displayed on a video monitor similar to a television set and is made up internally of dots (512×512) called pixels, each of which can have 256 (2^8) gray levels and is encoded to the image analyzer.

B. Software

It is the software included in the system that allows interpretation of the raw encoded data on pixel gray levels. By selecting a specific gray level, it is possible to highlight (digitize) those pixels at that level or the range of pixels between the selected gray level and the gray level opposite the background. Other digitizing schemes can also be employed. In the end, the digitized pixels make up the object for measurement.

In our system, plant embryos are highlighted through a range of gray levels, but along with the desired embryos are an array of objects that are not viable embryos. Further selection of the desired embryos is possible by measurement of the objects.

The software that is provided with the image analyzer (Image Technology Model 3000, Deer Park, New York) has a primitive shape analysis mode which is able to record eight types of measurement information on a digitized region that is a potential embryo observation to a file. This includes the position in Cartesian coordinates (the centroid), the area, the perimeter, the vertical diameter (height), the horizontal diameter (width), and the shortest (breadth) and longest (length) diameters of eight diameters taken at gradations of 22.5 degrees. These data can be calibrated to whatever units are desired.

III. REFINING THE IMAGING SOFTWARE

A standard procedure that enables the routine characterization of somatic embryos in a liquid culture by image analysis has been developed (Cazzulino, 1989). This was developed for carrot somatic embryos but is applicable to embryos of other species as well. The basic steps in the procedure are to determine measurement information (quantifiable parameters) that are best descriptive of somatic embryo morphology and compile statistical information based on these parameters for each somatic embryo developmental stage. This requires not only knowledge of normal somatic embryo development, but also a means for developing a one-to-one comparison between microscopic examination and image measurement of an embryo observation. The final step is then to develop an algorithm that assesses observations automatically by comparison to the statistical databank on normal somatic embryo development.

A. Parameter Selection

Quantifiable parameters are needed to distinguish embryo shape. The choice of a suitable set is not rigorous since many possibilities can be envisioned. These parameters were determined on the basis of the ability to give a quantitative description of embryo shape, the suitability for integration into the image analysis system, and an insensitivity to size and orientation.

Several sets of parameters have been considered for the characterization of shape. They include the coefficients in a so-called discrete Fourier transform (DFT) that describe the perimeter of an embryo (Cazzulino *et al.*, 1987), geometric parameters emphasizing descriptive but qualitative morphological characteristics based on Schiavone and Cooke (1985), and the available parameters that are incorporated into the image analysis software (Cazzulino *et al.*, 1990a).

The DFT parameter set has been found to be quite descriptive of shape (Cazzulino *et al.*, 1987). A DFT is defined as:

$$R(\Theta_k) = \sum_{n=1}^{N} C_n \exp(-in\Theta_k) \qquad \text{for k} = 1, \ldots M$$

where $R(\Theta_k)$ is the radial distance from the centroid to the perimeter at a particular angle Θ_k. C_n are the coefficients sought, where n is the harmonic counter and k is the angular counter. A DFT can have many coefficients, usually at multiples of 2, and the total number is chosen to give the best quantitative description for distinguishing different embryo shapes at minimum N. In fact, with enough coefficients (usually 16 or greater), an inverse DFT gives an accurate representation of the shape of the embryo originally measured (Cazzulino *et al.*, 1987). The difficulty with using DFTs is that very specific input data are needed to get the sought-after coefficients, C_n. The data are preferably equi-angularly spaced with reference to the centroid, and a multiple of 2. The operation of most image analyzers is not selective enough, does not have the necessary flexibility to find true centroids, and is not set up to report so-called primitive information on more than a limited number of parameters per observation. These factors discourage the implementation of a DFT parameter set.

Schiavone and Cooke (1985) devised geometric parameters to characterize the major stages of carrot embryogenesis. They are summarized by the lines $|ab|$, $|cd|$, and $|ef|$ in the following illustration:

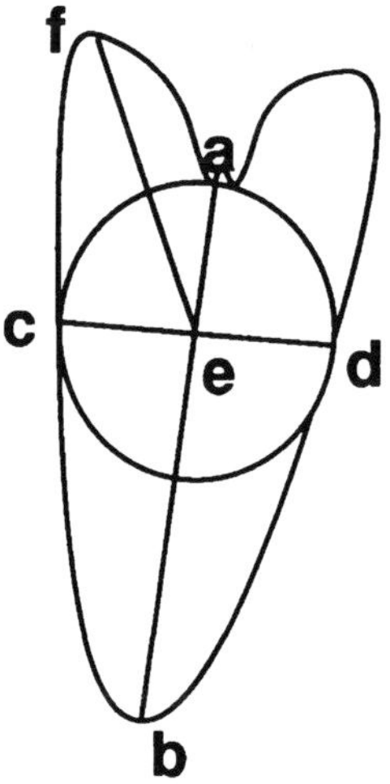

According to the authors, the lines represent the major growth vectors in developing embryos and were named the axial length $|ab|$, the embryo diameter $|cd|$, and the cotyledon arc $|ef|$. The advantage associated with using this technique is that the parameters are *specifically* designed to characterize the developmental stages of embryos similar to the qualitative descriptions appearing in the plant tissue culture literature. In other words, the parameters are chosen to register changes that specifically distinguish the developmental stages from a visual point of view. Embryo diameter is a measure of the increase in size from globular to torpedo. Axial length is a measure of the increase in elongation, and cotyledon arc is a measure of cotyledon formation and elongation. However, several complicating factors arise when this method is set up for the image analyzer. The orientation of the cotyledons may be such that point ***a*** is not observable and there is no easy way to find point ***e*** with the image analyzer, unless ***e*** as the centroid is a workable solution. Furthermore, even if point ***e*** is found, the cotyledon arc length may not be unambiguously determined. As a result, these parameters are not entirely appropriate for an image analysis system.

There are, however, four parameters that may typically be found directly from an image analyzer and are somewhat orientation independent. They are the area, A, the perimeter, P, and the shortest (breadth), B, and the longest (length), L, diameters of eight diameters taken at gradations of 22.5 degrees around the centroid. This parameter set was used as a framework for developing parameters that were both descriptive of geometric or morphological characteristics of embryos and easily available to most image analysis software (Cazzulino, 1989). In addition, this new parameter set can be made scale invariant, so as not to be dependent on dimensional terms, orientation, or magnification. Three invariant parameters that can be deduced from the four independent image analysis variables are:

$$E = \frac{L}{B}$$

$$R = \frac{P^2}{4\pi A} \quad \text{or} \quad \frac{2P}{\pi(L+B)}$$

$$F = \frac{P(L+B)}{8A}$$

E is a measure of "elongation," R is a measure of the "roughness" of the embryo outline, and F is a measure of "form." It is important to note that all of the parameters are independent.

The three invariant parameters were chosen because they emphasized geometric characteristics unique to single freely suspended embryos as they develop in culture medium. Cytoplasmically dense cell clusters, embryos stuck to each other, and deformed embryos are picked up as observations by the image analyzer, but because of convolutions in their shape and irregularities on their surfaces the roughness parameter can resolve them, and because of the lack of recognizable developmental organization the elongation parameter can separate them out. The characteristics of embryo development that distinguish the developmental stages are also emphasized by these parameters. For example, if the observation closely resembles a circle, is not elongated, and has a smooth outline, then the observation could be characterized as a globular embryo. Also, if the observation hardly resembles a circle, is elongated, and has a smooth but somewhat advanced developmentally organized outline, then the observation might be characterized as a torpedo embryo.

B. Distinguishing Shape by Statistical Methods

To distinguish and classify observations on an image analyzer, the parametric data have to be analyzed. Statistical procedures best suited for this type of analysis and the formulation of a shape recognition algorithm include multilevel simple statistical comparisons (MSSC) and classificatory discriminant analysis (CDA) (Cazzulino *et al.*, 1990a). The multilevel simple statistical comparisons have been used as a selection method for desired embryo observations (Cazzulino *et al.*, 1990a), and the classificatory discriminant analysis, an established statistical technique (Brugal *et al.*, 1986), was used to classify embryo observations into different stages of development or embryo classes. MSSC comparisons are between normal distributions of representative data and experimental data on an actual observation. The data on an observation are restricted to be within the boundaries of, say, two standard deviations of representative data for selection.

CDA, on the other hand, is a two-step multivariate procedure. First, from a matrix of representative statistical data for all the parameters on each embryo class, an axis (called a discriminant function) that most clearly distinguishes the classes is found. Each embryo class is designated a specific point along the axis from which a measure of how close an observation is to a class can be made. An observation is classified into the embryo class that is closest along this discriminant function axis.

C. Algorithm Development

The algorithm was developed by working backwards from the desired result. Hence, studies were first made on developing the classification system. Some of the steps necessary for accurate analysis include obtaining data on coded embryo observations, running the data through the trial algorithm, and then analyzing the accuracy of the results. This was usually followed by modifications to the algorithm.

A preliminary study of at least 20 embryos from each developmental stage followed these steps. Accuracy was assessed by comparing algorithm results to the encoded (i.e., known classification) microscope observation. A summary of the results is conveniently presented in a

TABLE I

Comparison of 61 Embryo Stage Classifications: The Imaging Algorithm versus Microscopic Observation

Number of microscopic observations for each embryo stage		Imaging algorithm classification by embryo stage of the same observations (percentage of microscopic observations)		
Embryo stage	Total	Globular	Heart	Torpedo
Globular	20	16 (80%)	4 (20%)	0 (0%)
Heart	21	4 (19%)	15 (71%)	2 (10%)
Torpedo	20	0 (0%)	4 (20%)	16 (80%)
Total	61	20	23	18

so-called confusion matrix as shown in Table I. Overall, about 77% agreement was found. Importantly, confusion occurred only among adjacent classes. That is, a globular embryo was never mistaken for a torpedo embryo, and vice versa. With approximately 80 observations in each class, the overall accuracy remained about the same (78%). When area was included as a parameter, the overall accuracy in discriminant classification improved to about 85% for the same observations, and accuracy in choosing torpedos was slightly over 95%. These percentages remained about the same when more than 1000 encoded observations were used, as demonstrated in Table II (Cazzulino, 1989).

These results indicate that discriminant classification with the imaging parameters can reliably give results similar to microscopic observation for the observations used to determine the discriminant function. This same reliability has to follow for observations that are not used to determine the discriminant function. When these observations were tested,

TABLE II

Comparison of More than 1000 Embryo Stage Classifications: The Imaging Algorithm versus Microscopic Observation

Number of microscopic observations for each embryo stage		Imaging algorithm classification by embryo stage of the same observations (percentage of microscopic observations)		
Embryo stage	Total	Globular	Heart	Torpedo
Globular	396	322 (81%)	74 (19%)	0 (0%)
Heart	358	99 (28%)	245 (68%)	14 (4%)
Torpedo	328	1 (0%)	43 (13%)	284 (87%)
Total	1082	422	362	298

TABLE III

Comparison of Embryo Stage Classifications Not Included in the Learning Set: The Imaging Algorithm versus Microscopic Observation

Number of microscopic observations for each embryo stage		Imaging algorithm classification by embryo stage of the same observations (percentage of microscopic observations)		
Embryo stage	Total	Globular	Heart	Torpedo
Globular	188	168 (89%)	20 (11%)	0 (0%)
Heart	154	75 (49%)	76 (49%)	3 (2%)
Torpedo	132	0 (0%)	8 (6%)	124 (94%)
Total	474	243	104	127

the overall accuracy for classifying 474 outside observations, listed in Table III, was more than 77%. In addition, these results show that the imaging system is very good at classifying torpedo (94%) and globular (89%) embryos but not heart embryos. This would be entirely expected, since heart embryos are a transitional stage.

These results are encouraging and show that discriminant classification works for outside observations as well. The method for analyzing the outside observations could also be extended to automatic operation.

The steps involved in developing the selection system were exactly the same as the classification system. In developing this part of the algorithm, the size and shape characteristics of a developing embryo were taken into account. That is, the algorithm was developed around the idea that the shape of a plant embryo was dependent on the size of the observed embryo. When a true embryo is small, it is spherical in shape, whereas when it is large, it is elongated. Hence, the elongation parameter was included in the algorithm to select small spherical, as well as large elongated, observations as embryos, and objects that did not fit these descriptions would not be selected as an embryo. An additional goal of the algorithm was to eliminate the selection of aberrant shapes that may approximate spherical or elongated structures but which have more convoluted forms. This was accomplished using the roughness parameter as a selection criteria.

The initial criteria for the selection system were based on 79 observations of globular embryos, 80 observations of heart embryos, and 81 observations of torpedo embryos. Another 138 observations, which did not include any of the 240 observations used for development of the criteria, were then run through the imaging system and compared with microscope evaluation. The results showed that there was 83% overall agree-

TABLE IV

Comparison of Embryo Identifications Compiled from Five Experimental Runs: The Imaging Algorithm versus Microscopic Observation

Number of microscopic observations		Imaging algorithm identification of observations (percentage of microscopic observations)	
Observation	Total	Embryo	Nonembryo
Embryo	425	397 (93%)	28 (7%)
Nonembryo	410	144 (35%)	266 (65%)
Total	835	541	294

ment between microscopic examination and the imaging system. After compilation of five different experimental runs, the overall agreement, listed in Table IV, was found to be closer to 79%. The overall agreement percentage was satisfactory, but the percentage of false positive identifications, which are observations identified by the imaging system to be embryos that were not considered embryos by microscopic observation, was unsatisfactory, at 35%. This could result in an overall agreement percentage as low as 65% if the majority of observations were not embryos. In addition, the selection of observations by the imaging system during the first stage of the algorithm that are not embryos is too strong a possibility, and improvements were made on the criteria for embryo selection through heuristic modification. The false positive identifications were reduced to 13% in this way, as shown in Table V.

The selection and classification systems were combined to form a two-step decision tree for embryo shape recognition that is schematically

TABLE V

Comparison of 1,758 Embryo Identifications: The Imaging Algorithm versus Microscopic Observation

Number of microscopic observations		Imaging algorithm identification of observations (percentage of microscopic observations)	
Observation	Total	Embryo	Nonembryo
Embryo	693	467 (67%)	226 (33%)
Nonembryo	1065	141 (13%)	924 (87%)
Total	1758	608	1150

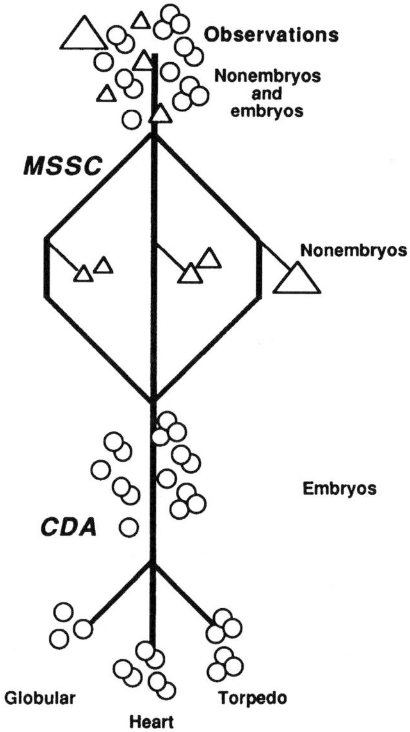

Fig. 1. A decision tree developed for the plant embryo image analysis system. Embryos (depicted as circle structures) are first distinguished from nonembryos (triangles) and then the developmental stage is determined.

summarized in Fig. 1. The initial step is the selection process in which an observation is classified into one of three groups and subsequently determined to be an embryo or not. Those observations that are selected as embryos are run through the embryo classification process for determination of appropriate embryo development stage.

IV. EMBRYO CONCENTRATION ASSESSMENT

It is possible to estimate the number concentration of embryos in each developmental stage in a sample by simply keeping track of how a sample is prepared and measured for image analysis. The embryo concentration data provide a quantitative assessment of an embryogenesis culture that is otherwise not possible. When conditions are changed in the culture, the effect on the embryos can be measured. This can lead to a rational method of improving embryo yield.

A. Factors Affecting Image Measurements

Sample preparation and image measurements are straightforward operations. Yet, several factors, such as magnification and dilutions, that are involved in these operations can affect embryo determinations. The choice of magnification is primarily influenced by the accuracy of measurements and the size of the measurement field.

The accuracy of measurements, in turn, is dependent on magnification because image analyzer data collection occurs in discrete units called pixels. Since a pixel is the minimum size measurement, the accuracy at any specified magnification is the pixel size. For instance, with the image analysis set-up in our laboratory, a magnification of $\times 25$, would measure a 100 μm globular embryo to an accuracy of $\pm 5\%$, whereas a magnification of $\times 15$, would measure the globular embryo to an accuracy of $\pm 9\%$.

The size of the measurement field increases with decreasing magnification. It would be best to maximize this variable for two reasons: first, to ensure that the largest expected object, such as a whole torpedo embryo, could fit in the field, and second, to maximize the number of observations that could be measured at a single time. An error as a result of embryos being only partially in the measurement field is another critical factor. At $\times 30$, in our laboratory, wrong embryo determinations could result in underestimates of torpedos by as much as 23%, whereas $\times 25$ results in underestimates of torpedos by about 15%, and uncertainties of $\pm 8\%$ and $\pm 4\%$ for heart and globular embryos, respectively. Therefore, the magnification that is best for measurement field size is worst for measurement accuracy. Hence, a compromise magnification has to be chosen.

Dilution of a sample is generally necessary because over time an embryogenic culture will produce a large number of embryos. Samples taken for image analysis must be sufficiently dilute so that there are only discrete and not overlapping objects in the imaging field. If objects touch each other, the tests outlined above would, in general, treat the objects as one observation, resulting in erroneous determinations. However, a large enough sample size has to be taken to assure an adequate representation of the culture. This can be resolved by correlating sample size to cell density (where sample size decreases as density increases), and diluting the samples to a constant volume.

B. Experimental Time Course

At a given interval in an embryogenic culture, a sample can be taken and the system can record the number of observations, number of embryos, and stage of each embryo per unit volume of culture. In addition, samples can be taken over the whole time course of the culture to get a quantitative assessment of overall embryo development. Hence, the system provides detailed quantitative descriptions of many facets of a culture that were not attainable previously. Several of these time course experiments are described below.

Initial time course experiments with carrot cell cultures were run in shake flasks (Cazzulino, 1989) to see if the time course of embryo development as monitored by the imaging system coincided with the generally accepted views on embryo ontogeny. The results obtained from the imaging system are shown in Fig. 2, and it is seen that globular and then heart embryos appear between 2 and 4 days after the initiation of the embryogenic culture. After 7 days the first torpedo embryos appear. This time period coincides with observations made by others (Nomura and Komamine, 1985; Warren and Fowler, 1978), although some reports indicate that globular embryos appear after 6–8 days, heart embryos after 9–10 days, and torpedo embryos only after as many as 12–14 days (Giuliano *et al.*, 1983; Schiavone and Cooke, 1985). These discrepancies could be the result of variations in the inoculum density, genotype, and culture conditions.

Total embryo concentrations were found to be as high as 6,000 embryos per ml around day 16 of the culture. These embryos have not been assessed for conversion frequency, however. Most literature on somatic embryogenesis does not contain information on the embryo concentration in culture, and qualitative information is all that is usually put forth. At least three groups, however, monitored embryo concentrations over time and assessed the developmental stage at each sampling period (Schiavone and Cooke, 1987; Sung *et al.*, 1979; Verma and Dougall, 1977). In these cases, only one or two time courses with this information were reported, which probably reflects the difficulty the researchers had in obtaining this information. The imaging system, on the other hand, can obtain the same information routinely.

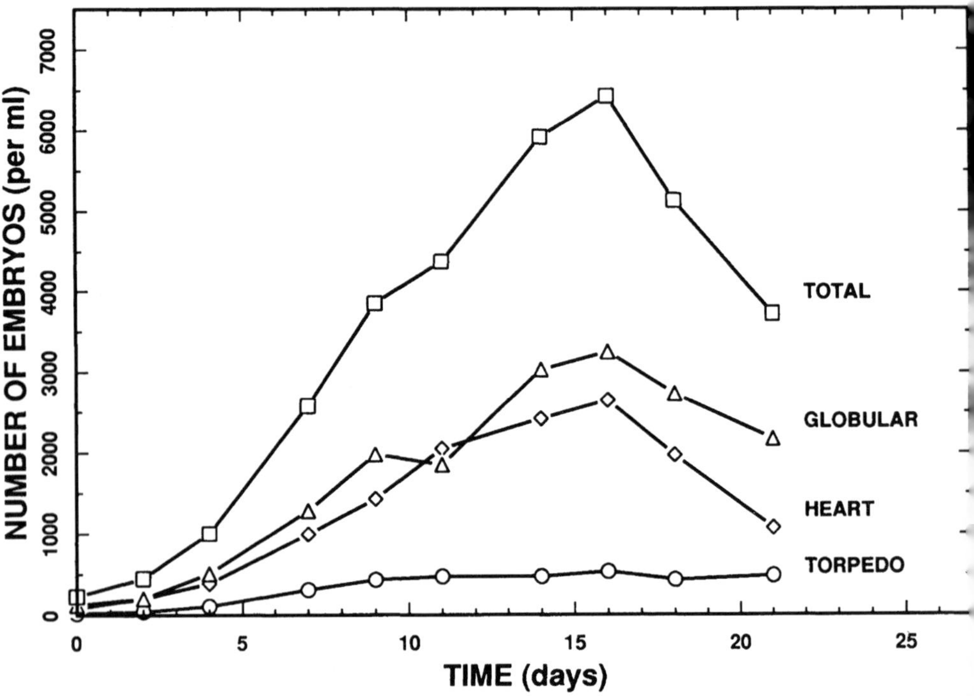

Fig. 2. The number concentration of carrot embryo developmental stages over the time course of an embryogenesis culture.

Different stages of development were expected to predominate at different periods over the time course as manifested by a sequential peaking of stages (Schiavone and Cooke, 1985; Verma and Dougall, 1977). This did not occur. Instead, all three stages of development peaked at the same time, roughly coinciding with the time extracellular carbohydrates were depleted. Throughout the time course illustrated in Fig. 2, the largest proportion of embryos were globular. Heart embryos formed an intermediate population size and torpedo embryos made up the smallest proportion. The concentration of both the globular and heart embryos fell off rapidly, while the torpedo embryos remained at a steady concentration. The fall off coincided with carbohydrate depletion. The total number of observations (embryos and nonembryos) fell off also, but to a lesser extent. The rapid fall off of the globular and heart embryo number densities could be the result of lysed cells that may cause small embryos and aggregates to stick together by agglomerating to both structures, resulting in nonembryo determinations which previously were embryo and nonembryo determinations.

Since early accounts on somatic embryogenesis, aberrant embryos have been described (Halperin and Wetherell, 1964). These may result from improper hormonal balance (Ammirato, 1987), incorrect lighting conditions (Michler and Lineberger, 1987), or extremely low cell density (Halperin, 1967; Warren and Fowler, 1981). The percentage of aberrant embryos as part of the total embryo number varies as a result of these conditions. With our image analysis system, later stage torpedo embryos, aggregates, and aberrant embryos are grouped into one class called nonembryos. They have been found to comprise anywhere from 85% to 40% of the observations in at least 7-day old embryogenic cultures. These percentages are in the same range as that of Halperin (1967), who estimated the percentage of embryos that developed from aggregates at 20%.

The stock cell culture over the period of 9 months lost some of its embryogenic potential. Total embryo formation decreased from more than 6000 embryos per ml in the early studies to slightly more than 2000 embryos per ml in the later studies. This phenomenon has been reported earlier for somatic embryogenesis (Reinert, 1970; Smith and Street, 1974). In light of this, studies analyzing modifications in experimental conditions should include control studies using the same age suspension cell culture.

C. Auxin Effects on Embryo Development

Numerous reports have shown that auxins inhibit embryo development (Fujimura and Komamine, 1979; Reinert, 1958; Schiavone and Cooke, 1987). Depending on the concentration or type of auxin, complete inhibition, delayed development, or partial inhibition of embryo development occurs. For example, 2,4-dichlorophenoxyacetic acid (2,4-D) is considered a strong inhibitor, whereas the natural auxin indole-3-acetic acid (IAA) is a weaker inhibitor.

We have carried out experiments that use the imaging system as a method to quantify these qualitative observations (Cazzulino, 1989). Three embryogenic cultures were run under the same experimental conditions except that the first culture contained no exogenous auxin, the second contained 0.05 μM of 2,4-D (that is 10% of the concentration in the proliferation culture), and the third contained 0.5 μM of IAA. For all three cultures, growth as measured by increase in dry weight was comparable. The doubling time for all three cultures was about 4 days, and the yields ranged from 0.29 to 0.32 g of cells per gram of carbohydrate utilized. The carbohydrates in the medium were depleted in the same pattern and at about the same time as described in the previous section.

The time course comparison of total embryo formation between the three experimental runs is shown in Fig. 3 and indicates that inhibition of embryo formation occurred in the culture with 2,4-D but not in the culture with IAA, even though the concentration of IAA was higher. Up to day 10, all three cultures produced similar amounts of embryos. After day 10, the 2,4-D culture contained lower concentrations of embryos. In addition, the peak concentration of embryos in 2,4-D medium occurred earlier and was about half that of the other cultures. The inhibition by 2,4-D was more pronounced for later stage embryos. For instance, globular concentrations were similar in all three cultures up until day 13. Torpedo formation was apparent on day 10 for the IAA culture and the auxin-free culture but was not apparent with the 2,4-D culture until day 18, which happened to be the only sampling period with any noticeable torpedo concentration.

Embryo development for the auxin-free culture and the IAA culture

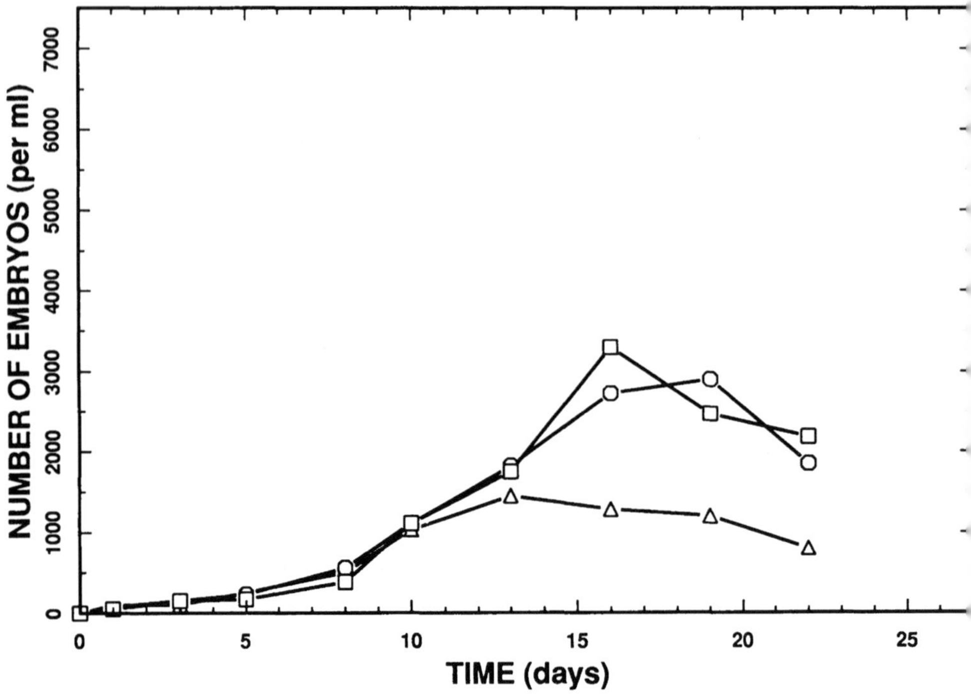

Fig. 3. The total number concentration for carrot embryo developmental stages over the time course of embryogenesis cultures containing various growth regulators. □, no auxin; ○, IAA; △, 2,4-D.

was similar. The globular embryo concentration peaked at the same time (day 16) and at a similar level (over 2,000 embryos/ml). Heart embryo concentrations peaked at a similar level of just under 1,000 embryos/ml, though they did not peak at the same time. Torpedo embryos appeared in culture at the same time and mostly maintained levels at around 200 embryos/ml. The proportions between embryo stages are similar to the data presented in Fig. 2, though at lesser levels.

V. BIOREACTOR OPERATION

In order to get a better understanding of the occurrences happening in an embryogenic culture and in order to develop a rational approach to testing the various chemical and environmental conditions for proper scale-up and operation of prototype bioreactors, a kinetic model is needed (Cazzulino *et al.*, 1990b). Such a model defines the important parameters in the operation of reactors for embryo production and can be used to test reactor operating strategies. It also forms the framework on which to compare the results of different experiments dealing with hormone levels, nutrient conditions, dissolved gases, temperature, or pH, among others. The following proposed model is based on time course data made possible by the imaging system. It is shown as an illustration of potential improvements in analysis and design made possible by the imaging system and not as the definitive mathematical description of an embryogenic culture.

A. Scale-Up Parameters

A simple schematic is shown in Fig. 4 that presents the developmental events occurring in an embryogenic culture as discrete compartments consisting of inoculum, preglobular, globular, heart, torpedo, aberrant embryo, and nonembryo types. The embryo proceeds through these stages until the end of the torpedo stage, where, when both the radicle and cotyledons are clearly discernible, the embryo is considered a plantlet. In the model, transition kinetics are assumed to be irreversible and are the result of growth, normal development, abnormal development, and death. This movement is represented by arrows in the schematic, with the thicker arrows indicating qualitatively higher rates.

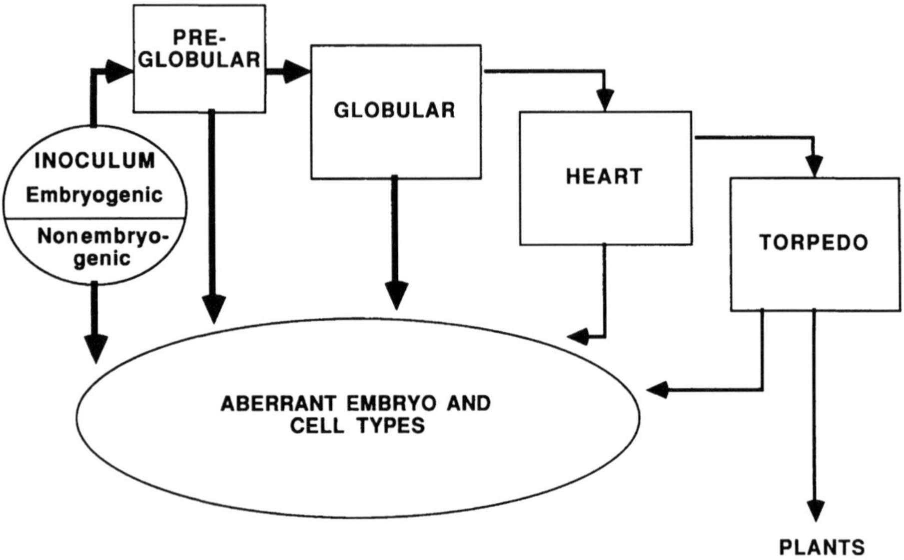

Fig. 4. A schematic representation of a model for plant somatic embryo development in an embryogenic culture.

The mass balance for the change in biomass, x, within a compartment i, can be expressed generally as:

$$\frac{dx_i}{dt} = \mu_i x_i - k_i x_i S + k_{i-1} x_{i-1} S - \kappa_i x_i S - k_{id} x_i; \quad i = 1, 2, \ldots, 5$$

where t is time, μ is the specific growth rate, k is a transition rate constant that is due to morphological development, κ is the aberrant embryo or cell rate constant, S is the substrate concentration, and k_{id} is the specific death rate. On the right-hand side, the first three terms represent the growth rate, the transition to further developmental stages, and the transitions from previous developmental stages. The last two terms represent the formation of aberrant cell or embryo types and cell or embryo death. The rate constants are expected to be dependent on substrate concentration, but the exact dependence is unknown, and for the sake of simplicity, these dependencies can be assumed to be first order. The specific growth rate, on the other hand, is known to be dependent on substrate concentration in a more structured way. For example, when sucrose is used as the carbohydrate source in carrot cell suspensions, it is hydrolyzed in the media (Edelman and Hanson, 1971; Fowler, 1982; Kanabus *et al.*, 1986), and only the hexoses are taken up into the cells. Furthermore, competitive inhibition is observed in the uptake rate of fructose (Bieleski, 1962; Fowler and Stepan-Sarkissian, 1985; Kanabus *et al.*, 1986). Hence, the following equation is used for μ_i;

$$\mu_i = \frac{\mu_{img} S_g}{(S_g + K_g)} + \frac{\mu_{imf} S_f}{(S_f + K_f(1 + S_g/K_g))}; \quad i = 1, 2, \ldots, 6$$

where K is a Monod type constant, μ_m is the maximum growth rate, and the subscripts g and f designate glucose and fructose, respectively. The substrate utilization mass balances include:

$$\frac{dS_s}{dt} = -k_s S_s$$

$$\frac{dS_g}{dt} = 0.526 k_s S_s - \frac{1}{Y_{xg}} \left(\sum_{i=1}^{6} \frac{\mu_{img} S_g}{S_g + K_g} x_i \right)$$

$$\frac{dS_f}{dt} = 0.526 k_s S_s - \frac{1}{Y_{xf}} \left(\sum_{i=1}^{6} \frac{\mu_{imf} S_f}{S_f + K_f(1 + S_g/K_g)} x_i \right)$$

The factor 0.526 reflects the mass change in the reaction of sucrose with water to yield glucose and fructose. Cell death during early embryo development (up to preglobular) is not important under our experimental conditions. At later developmental stages, embryo death results in the breakup of the defined structure into pieces of "undefined" structures. These still have cell mass and, as a result, were treated as elements transferred to the aberrant embryo compartment. This results in the following mass balance for the aberrant embryo and cell compartment:

$$\frac{dx_6}{dt} = \mu_6 x_6 - \sum_{i=1}^{5} k_i x_i S + \sum_{i=3}^{5} k_{id} x_i - k_{6d} x_6$$

The parameter values for these equations are listed in Table VI and are found from our experimental results with supporting information from Bieleski (1962), Fienberg *et al.* (1984), Fowler (1982), Fowler and Stepan-Sarkissian (1985), Fujimura and Komamine (1980), Kanabus *et al.* (1986), and Warren and Fowler (1978). For a given set of initial conditions, the system of equations can be solved using standard techniques.

The model uses the initial conditions corresponding to the results in Fig. 2. The values are listed in Table VII. Figure 5 presents the simulation results, which are similar to the experimental data depicted in Fig. 2. The embryo numbers increase at about the same time and rate for each developmental stage and peak at approximately the same time but do not fall off as rapidly as in the experiment. The reason behind this rapid drop-off in embryo number is unclear. However, it is not unusual to consider that the death rate is carbohydrate dependent since at low sugar

TABLE VI
The Parameter Values for the Carrot Somatic Embryo Model

Parameter	Dimension	Value	Reference
μ_{1m}	day^{-1}	0.1	Fienberg *et al.*, 1984
μ_{2m}		0.67	
μ_{3m}		0.18	
μ_{4m}		0.18	
μ_{5m}		0.18	
μ_{6m}		0.25	
k_{1d}	day^{-1}	0.015	Experimental estimate
k_{2d}		0.015	
k_{3d}		0.035	
k_{4d}		0.035	
k_{5d}		0.01	
k_{6d}		0.01	
k_1	liter/g day	0.003	Fujimura and Komamine,
k_2		0.02	1980; Warren and Fowler, 1978;
k_3		0.007	and experimental estimate
k_4		0.005	
k_5		0.01	
κ_1	liter/g day	0.0001	Experimental estimate
κ_2		0.0005	
κ_3		0.0005	
κ_4		0.0005	
κ_5		0.0001	
k_s	day^{-1}	1.2	Kanabus *et al.*, 1986; and experimental estimate
K_g	g/liter	2	Bieleski, 1962; Fowler, 1982
K_f	g/liter	13	Kanabus *et al.*, 1986; and experimental estimate
Y_{xg}	g/g	0.32	Experimental estimate
Y_{xf}	g/g	0.32	Experimental estimate

TABLE VII
Initial Conditions for the Model Simulations

Concentration variable	Value (g/liter)
Inoculum concentration	
Embryogenic	1.594
Nonembryogenic	0.176
Sucrose concentration	36.2
Glucose concentration	3.2
Fructose concentration	3.2

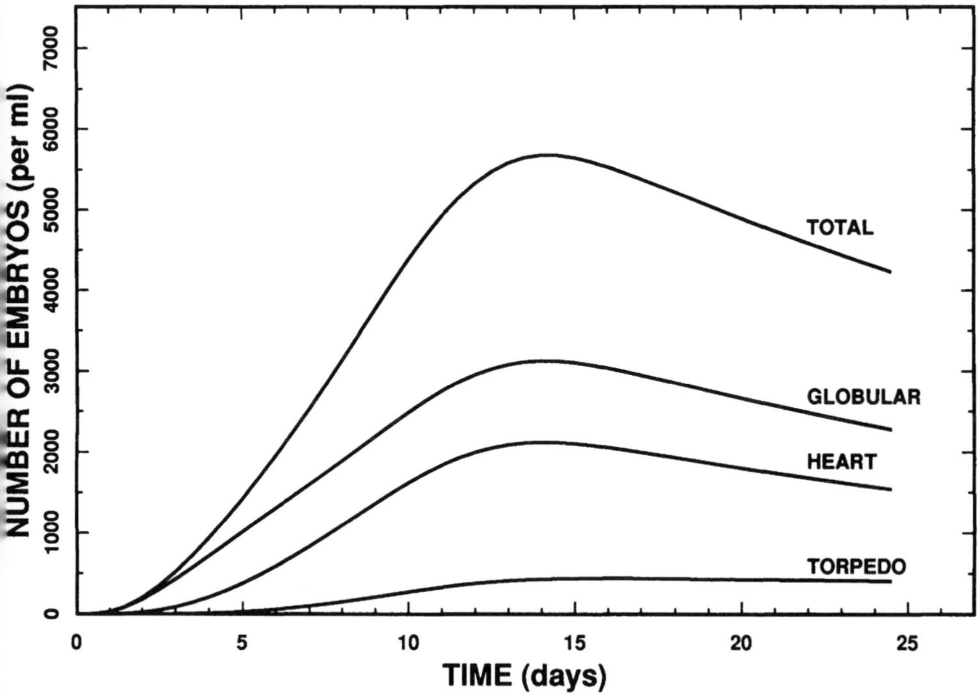

Fig. 5. Model simulation of the concentration of different carrot embryo stages corresponding to the conditions shown in Fig. 2.

concentrations the death of embryos increases. The model currently does not take this into account. Other factors may also come into play.

Certain factors not explicitly included in the model have a major effect on the rate of embryogenesis, such as the concentration of 2,4-D in the initiation suspension culture, the dissolved oxygen levels, and the age of the culture. All these factors affect the percentage of biomass that is embryogenic. This percentage is a variable in the model, which can be changed. With a better understanding of the effects of 2,4-D, dissolved gases, and culture age, a better representative model can be developed.

Hence, the model as it stands now would not be able to predict results under all conditions. However, it establishes a rational approach to understanding the factors affecting embryo development in liquid culture by clearly pointing out what rate constants or rate expressions are associated with different environmental factors. In this way, the model should help in scale-up and in optimizing the bioreactor production of mature embryos capable of developing into seedlings.

B. Bioreactor Types

Most reviews and research reports on bioreactors for plant cells are concerned with the production of plant secondary metabolites as the key product (see for example Kargi and Rosenberg, 1987). There have been only a few reports on the use of bioreactors for large-scale propagation of plants (Ammirato and Styer, 1985; Asenjo and O'Carroll, 1988; Chen *et al.*, 1987; Kessell and Carr, 1972; Preil *et al.*, 1988; Stuart *et al.*, 1987; Styer, 1985; Takayama *et al.*, 1986; Wheat *et al.*, 1986). Some of the major differences between reactors designed for metabolites and reactors designed for production of plant tissue include, in the latter case, the necessity to deal with media manipulations, the relatively large size of tissue structures in the reactor, and, in contrast, the relatively small anticipated scale of an eventual production system. For example, media changes are absolutely essential to the coordinated development of embryos from suspension cell cultures. In contrast, media manipulations that require complete modification of the culture conditions are rarely used in chemical production schemes. Furthermore, a 500 μm diameter embryo structure is almost 1000-fold larger in volume than a 50 μm diameter cell or cell aggregate that is typical of suspension cultures. The mixing requirements of the reactor, the monitoring of the reactor culture, and the handling of the reactor contents can be expected to be quite different under these types of conditions. Finally, the production sizes that would be appropriate for the manufacture of plant secondary metabolites could reasonably be estimated at around the 1000 liter scale, whereas "large" for an embryo production process might be on the order of 10 liters. The small size is dependent on the need for various types of media changes and can be an advantage since its monitoring and operation are likely to be easier to maintain under aseptic conditions.

Reactor configurations that have been tested for plant embryo production are summarized in Table VIII. In all cases, relatively small reactors were employed. The predominant class of reactor used is the suspension culture reactor that provides a uniform culture environment and allows for the easy control and regulation of operating conditions. The suspension culture reactor can be subdivided into mechanically agitated and nonmechanically agitated reactors. The mechanically agitated reactors can be driven by impellers, magnetic stir bars, or vibrating perforated plates. Since feedstock cells for embryo production are the same cells used in other plant cell cultures, concern for shear effects on cell viability has to be taken into account. Hence, in impeller systems a paddle type impeller is generally better than flat-blade turbine impellers (Tanaka, 1981). Vibromixed perforated plates (Chen *et al.*, 1987) and magnetically stirred systems (Chen *et al.*, 1987; Styer, 1985) have also proved effective.

An example of a nonmechanically agitated reactor that has been used in plant propagation studies is the airlift fermenter (Chen *et al.*, 1987; Stuart *et al.*, 1987). Since there is usually less shear generated in an airlift, loss of cell viability is not a typical problem in this type of system. However, the range of usable aeration rates can be restrictive (Ammirato and Styer, 1985), and air stripping of essential nutrients or product gases needed for embryogenesis may arise (Chen *et al.*, 1987). Nevertheless, Stuart *et al.* (1987) found that the production of alfalfa embryos in airlift systems can be greater than that found in shake flasks under control conditions.

Several authors advocate the use of magnetically stirred fermenters (Ammirato and Styer, 1985; Chen *et al.*, 1987; Styer, 1985). Compared with other reactor configurations, Chen *et al.* (1987) obtained high embryo yields with this configuration. A modified magnetically stirred reactor called a spin-filter reactor seems to be especially suitable for bioreactor plant propagation (Styer, 1985) and is the only reactor designed specifically for media changes or staged operations. A schematic diagram of this type of reactor is shown in Fig. 6. The spin-filter bioreactor has a central shaft that supports a housing for a filter that is attached to the magnetic stirrer. The housing and stirrer rotate but not the shaft. The shaft is hollow and serves as the means for removal of medium withdrawn through the spinning filter. This design allows the exchange of medium without cell loss, which can be exploited when changing medium to induce embryogenesis. Carrot somatic embryos have been successfully produced by this method (Ammirato and Styer, 1985).

Also listed in Table VIII are the conversion frequencies for some of the bioreactor types that have been used with embryogenic cell cultures. In general, the conversion of embryo structures into plants with normal morphology is a limitation of current bioreactor operations compared with cultures grown in shake flask systems (Stuart *et al.*, 1987). In the experiments shown in Table VIII with cultures of poinsettia (Preil *et al.*, 1988), embryos were counted based on the conversion frequency alone. In addition, these cultures were further stimulated to produce embryos after bioreactor growth by plating onto embryogenesis media, and the results may therefore appear unnaturally high. Nevertheless, the results are encouraging for the development of effective bioreactor systems.

It is also apparent from the list shown in Table VIII that very few plant species have been investigated for somatic embryogenesis in bioreactors. In view of this, it is likely that there is considerable room for improvement of operating strategies and final embryo yields. The yields are, in fact, much lower than can be found in "optimized" shake flask cultures where embryo number densities (counted as total structures) can be as high as several thousand (see Fig. 2). One of the parameters

TABLE VIII
Bioreactor Designs Used for Scale-Up of Embryo Production

Plant species	Reactor size and type, design, and production time[a]	Embryo production[b] (number/ml)	Dissolved oxygen (DO) levels (aeration rate[c])	Other information	Reference
Carrot (*Daucus carota*)	4 liter stirred tank (in-house design) 60 days	<1 1–10	>16% (0.5 vvm) <16% (0.025 vvm)	Higher DO levels were obtained by increasing direct sparging as well as mechanical agitation; rhizogenesis observed at high DO levels	Kessel and Carr (1972)
Carrot	<1 liter spin filter (in-house design) n/a	Conversion frequency[d] reported to be comparable to controls	n/a	Two-stage reactor design	Ammirato and Styer (1985); Styer (1985); Wheat *et al.* (1988)
Alfalfa (*Medicago sativa*)	1.6 liter airlift reactor (commercially available) n/a	0	100% (0.5 vvm)	Examined at high DO levels only	Chen *et al.* (1987)
	1 liter hanging stir-bar fermentor (HSBF) (commercially available) n/a	9–10 (30–80% conversion frequency)	50% (average value)	At low DO levels (<10%) the number of embryos formed falls off; control cultures contained 30–44 embryos/ml	

Alfalfa	1 liter stirred tank (commercially available) 14 days	0	21% (0 vvm)	Control cultures contained 139 embryos/ml	Stuart *et al.* (1987)
		80 (2–3% conversion frequency)	70% (1–2 vvm)		
Poinsettia (*Euphorbia pulcherrima*)	2 liter stirred tank (commercially available) 28 days	100 (based on counting only embryos with 100% conversion frequency)	60%	Silicone tubing used to avoid direct sparging; agitation achieved with a Vibromixer™; After harvesting, filtrates were grown 4 additional weeks on agar media for embryo counts.	Preil *et al.* (1988)

[a] n/a indicates no values reported.
[b] Embryo concentrations are reported as embryo "structures" observed at the end of the reactor run unless otherwise indicated.
[c] Aeration rates are volumes of air sparged per volume of liquid in the reactor per minute (vvm).
[d] Conversion frequencies are measured by counting the fraction of embryo "structures" that germinate into plants.

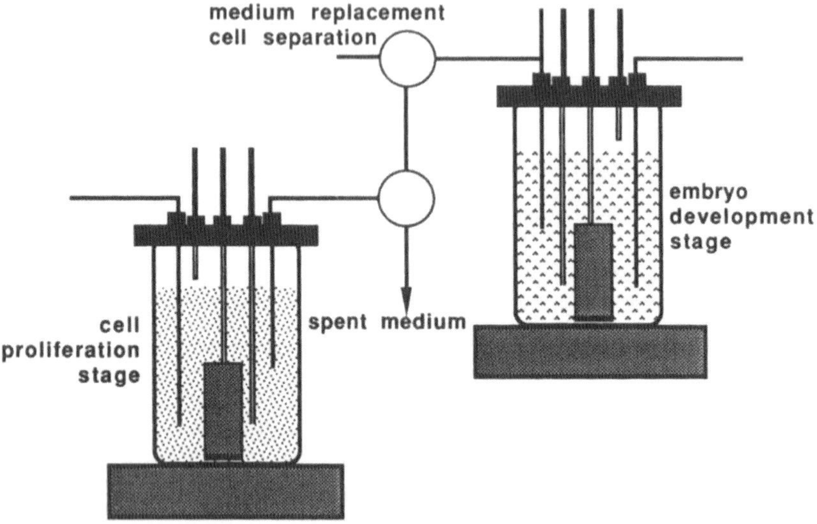

Fig. 6. Staged reactor design in a spin-filter system that allows for media manipulation and independent control over cell proliferation and embryogenic development.

investigated in bioreactors that affects the yield is the dissolved oxygen (DO) concentration (Kessell and Carr, 1972; Stuart *et al.*, 1987; Preil *et al.*, 1988). For *Daucus carota* cultures, low DO concentration is conducive to formation of embryogenic structures, while for *Medicago sativa* conflicting reports have been published. In sparged vessels, other gases besides oxygen, such as carbon dioxide and ethylene, may be modulated. It remains to be seen what are the true effects of DO on embryogenic potential.

VI. CONCLUSION

Major strides have recently been made toward establishing somatic embryogenesis as a viable plant micropropagation technique. Carrot somatic embryo structures in a suspension culture can now be reliably and routinely quantified. This provides a means toward improving conditions for increasing embryo production in culture using sophisticated bioreactors. Furthermore, a mathematical model has been proposed in order to give a more rational approach to finding the keys to unlocking the potential of increasing embryo production. These achievements lay the groundwork for additional research on improving bioreactor cloning systems and on developing a fully automated process that couples bioreactor operation, an imaging system, an embryo separation process, and a downstream delivery system.

REFERENCES

Aitken-Christie, J., Singh, A. P., and Davies, H. (1987). Multiplication of meristemic tissue—a new tissue culture system for radiata pine. *In* "Genetic Manipulation of Woody Plants" (J. Hanover, and D. Keathley, eds.), pp. 413–432. Plenum Press, New York.

Ammirato, P. V. (1987). Organization events during somatic embryogenesis. *In* "Plant Tissue and Cell Culture" (C. E. Green, D. A. Somers, W. P. Hackett, and D. D. Biesboer, eds.), pp. 57–81. Alan R. Liss, New York.

Ammirato, P. V., and Styer, D. J. (1985). Strategies for large scale manipulation of somatic embryos in suspension culture. *In* "Biotechnology in Plant Science: Relevance to Agriculture in the Eighties" (M. Zaitlin, P. Day, and A. Hollaender, eds.), pp. 161–178. Academic Press, New York.

Asenjo, J. A., and O'Carroll, P. J. (1988). Physiological and biochemical engineering factors in whole plant regeneration from *in-vitro* cell suspension cultures. *196th ACS National Meeting*. Los Angeles, CA. paper no. 110. *Abstract.*

Backs-Hüsemann, D., and Reinert, J. (1970). Embryobildung durch isolierte Einzelzellen aus Gewebekuturen von *Daucus carota*. *Protoplasma* **70,** 49–60.

Barnett, C. E., White, R. A., Petrovic, A. M., and Good, G. L. (1987). An automated apparatus for measuring root length. *HortScience* **22,** 140–144.

Bieleski, R. L. (1962). The physiology of sugar-cane V. kinetics of sugar accumulation. *Aust. J. Biol. Sci.* **15,** 429–444.

Brugal, G., Quirion, C., and Vassilakos, P. (1986). Detection of bladder cancers using a SAMBA-200 image processor. *Anal. Quant. Cytol. Histol.* **8,** 187–192.

Cazzulino, D. (1989). Plant somatic embryos in culture: quantification and identification by image analysis, and embryogenesis culture kinetics. Ph.D. Chemical and Biochemical Engineering. Rutgers, The State University of New Jersey.

Cazzulino, D., Pedersen, H., and Chin, C.-K. (1990a). Monitoring somatic embryo development in carrot cell cultures. *Dev. Ind. Microbiol.* **31,** 285–292.

Cazzulino, D., Pedersen, H., Chin, C.-K., and Styer, D. (1990b). Kinetics of carrot somatic embryo development in suspension culture. *Biotechnol. Bioeng.* **35,** 781–786.

Cazzulino, D., Pedersen, H., Chin, C.-K., Venkat, K., and Styer, C. (1987). Characterization of plant somatic embryo development using Fourier shape analysis. *Ann. New York Acad. Sci.* **506,** 190–195.

Chen, T. H. H., Thompson, B. G., and Gerson, D. F. (1987). *In vitro* production of alfalfa somatic embryos in fermentation systems. *J. Ferment. Technol.* **65,** 353–357.

Edelman, J., and Hanson, A. D. (1971). Sucrose suppression of chlorophyll synthesis in carrot tissue cultures: The role of invertase. *Planta* **101,** 122–132.

Fienberg, A. A., Choi, J. H., Lubich, W. P., and Sung, Z. R. (1984). Developmental regulation of polyamine metabolism in growth and differentiation of carrot culture. *Planta* **162,** 532–539.

Fowler, M. W. (1982). Substrate utilisation by plant–cell cultures. *J. Chem. Technol. Biotechnol.* **32,** 338–346.

Fowler, M. W., and Stepan-Sarkissian, G. (1985). Carbohydrate source, biomass productivity and natural product yield in cell suspension cultures. *In* "Primary and Secondary Metabolism of Plant Cell Cultures" (K. H. Neumann, W. Barz, and E. Reinhard, eds.), pp. 66–73. Springer-Verlag, Berlin.

Fujimura, T., and Komamine, A. (1979). Synchronization of somatic embryogenesis in a carrot cell suspension culture. *Plant Physiol.* **64,** 162–164.

Fujimura, T., and Komamine, A. (1980). The serial observation of embryogenesis in a carrot cell suspension culture. *New Phytolologist* **86,** 213–218.

Giuliano, G., Rosellini, D., and Terzi, M. (1983). A new method for the purification of the different stages of carrot embryoids. *Plant Cell Rep.* **2,** 216–218.

Grand d'Esnon, A. G., Chee, R., Harrell, R. C., and Cantliffe, D. J. (1988). Qualitative and Quantitative Evaluation of Liquid Tissue Culture by Artificial Vision. *Agricultural Engineering International Conference.* Paris. paper no. 395.

Halperin, W. (1967). Population density effects on embryogenesis in carrot-cell cultures. *Experimental Cell Res.* **48,** 170–173.

Halperin, W., and Wetherell, D. F. (1964). Adventive embryony in tissue cultures of the wild carrot, *Daucus carota. Am. J. Bot.* **51,** 274–283.

Kanabus, J., Bressan, R. A., and Carpita, N. C. (1986). Carbon assimilation in carrot cells in liquid culture. *Plant Physiol.* **82,** 363–368.

Kargi, F., and Rosenberg, M. Z. (1987). Plant cell bioreactors: Present status and future trends. *Biotechnol. Prog.* **3,** 1–8.

Kessell, R. H. J., and Carr, A. H. (1972). The effect of dissolved oxygen concentration on growth and differentiation of carrot (*Daucus carota*) tissue. *J. Exp. Bot.* **23,** 996–1007.

Lutz, J. D., Wong, J. R., Rowe, J., Tricoli, D. M., and Lawrence, J., R.H. (1985). Somatic embryogenesis for mass cloning of crop plants. *In* "Tissue Culture in Forestry and Agriculture" (R. R. Henke, K. W. Hughes, M. J. Constantin, and A. Hollaender, eds.), pp. 105–116. Plenum Publishing, New York.

Michler, C. H., and Lineberger, R. D. (1987). Effects of light on somatic embryo development and abscisic levels in carrot suspension cultures. *Plant Cell Tissue Organ Cult.* **11,** 189–207.

Nomura, K., and Komamine, A. (1985). Identification and isolation of single cells that produce somatic embryos at a high frequency in a carrot suspension culture. *Plant Physiol.* **79,** 988–991.

Preil, W., Florek, P., Wix, U., and Beck, A. (1988). Towards mass propagation by use of bioreactors. *Acta Hortic.,* **226,** 99–106.

Redenbaugh, K., Paasch, B. D., Nichol, J. W., Kossler, M. E., Viss, P. R., and Walker, K. A. (1986). Somatic seeds: Encapsulation of asexual plant embryos. *Bio/Technol.* **4,** 797–801.

Redenbaugh, K., Viss, P., Slade, D., and Fujii, J. (1987). Scale-up: Artificial seeds. *In* "Plant Tissue and Cell Culture" (C. E. Green, D. A. Somers, W. P. Hackett, and D. D. Biesboer, eds.), pp. 473–493. Alan R. Liss, New York.

Reinert, J. (1958). Untersuchungen uber die morphogenese an gewebekulturen. *Ber. Dtsch. Bot. Ges.* **71,** 15.

Reinert, J. (1970). Determination of embryo and root formation in tissue cultures from *Daucus carota. In* "Les Cultures de Tissus de Plantes" (M. L. Hirth, and M. G. Morel, eds.), pp. 261–268. Centre National de la Recherche Scientifique, Paris.

Schiavone, F. M., and Cooke, T. J. (1985). A geometric analysis of somatic embryo formation in carrot cell cultures. *Can. J. Bot.* **63,** 1573–1578.

Schiavone, F. M., and Cooke, T. J. (1987). Unusual patterns of somatic embryogenesis in the domesticated carrot: Developmental effects of exogenous auxins and auxin transport inhibitors. *Cell Diff.* **21,** 53–62.

Smith, M. A. L., and Spomer, L. A. (1987). Direct quantification of *in vitro* cell growth through image analysis. *In Vitro Cell. Dev. Biol.* **23,** 67–74.

Smith, M. A. L., Spomer, L. A., Meyer, M. J., and McClelland, M. T. (1989). Non-invasive image analysis evaluation of growth during plant micropropagation. *Plant Cell, Tissue Organ Cult.* **19,** 91–102.

Smith, S. M., and Street, H. E. (1974). The decline of embryogenic potential as callus and suspension cultures of carrot (*Daucus carota* L.) are serially subcultured. *Ann. Bot.* **38,** 223–241.

Steward, F. C., Mapes, M. O., and Mears, K. (1958). Growth and organized development of cultured cells, II. organization in cultures grown from freely suspended cells. *Am. J. Bot.* **45,** 704–708.

Stuart, D. A., Nelsen, J., McCall, C. M., Strickland, S. G., and Walker, K. A. (1985a). Physiology of the development of somatic embryos in cell cultures of alfalfa and celery. *In* "Biotechnology in Plant Science: Relevance to Agriculture in the Eighties" (M. Zaitlin, P. Day, and A. Hollaender, eds.), pp. 35–47. Academic Press, New York.

Stuart, D. A., Nelsen, J., Strickland, S. G., and Nichol, J. W. (1985b). Factors affecting developmental processes in alfalfa cell cultures. *In* "Tissue Culture in Forestry and Agriculture" (R. R. Henke, K. W. Hughes, M. J. Constantin, and A. Hollaender, eds.), pp. 59–73. Plenum Publishing, New York.

Stuart, D. A., Strickland, S. G., and Walker, K. A. (1987). Bioreactor production of alfalfa somatic embryos. *HortScience* **22,** 800–803.

Styer, D. J. (1985). Bioreactor technology for plant propagation. *In* "Tissue Culture in Forestry and Agriculture" (R. R. Henke, K. W. Hughes, M. J. Constantin, and A. Hollaender, eds.), pp. 117–130. Plenum Publishing, New York.

Sung, Z. R., Smith, R., and Horowitz, J. (1979). Quantitative studies of embryogenesis in normal and 5-methyltryptophan-resistant cell lines of wild carrot. *Planta* **147,** 236–240.

Takayama, S., Arima, Y., and Akita, M. (1986). Mass propagation of plants by fermentor culture techniques. *In* "Abstracts VI International Congress of Plant Tissue and Cell Culture" (D. A. Somers, B. G. Gegenbach, D. D. Biersboer, W. P. Hackett, and C. E. Green, eds.), pp. 449. International Association of Plant Tissue Culture, Minneapolis.

Tanaka, H. (1981). Technological problems in cultivation of plant cells at high density. *Biotechnol. Bioeng.* **23,** 1203–1218.

Verma, D. C., and Dougall, D. K. (1977). Influence of carbohydrates on quantitative aspects of growth and embryo formation in wild carrot suspension cultures. *Plant Physiol.* **59,** 81–85.

Warren, G. S., and Fowler, M. W. (1978). Cell number and cell doubling times during the development of carrot embryoids in suspension culture. *Experientia* **34,** 356–357.

Warren, G. S., and Fowler, M. W. (1981). Physiological interactions during the initial stages of embryogenesis in cultures of *Daucus carota* L. *New Phytologist* **87,** 481–486.

Wheat, D., Bondaryk, R. P., and Nystrom, J. (1986). Spin filter bioreactor technology as applied to industrial plant propagation. *HortScience* **21,** 819.

Zayas, I., Lai, S. S., and Pomeranz, Y. (1986). Discrimination between wheat classes and varieties by image analysis. *Cereal Chem.* **63,** 52–56.

CHAPTER 9

Automated Evaluation of Somatic Embryogenesis in Sweet Potato by Machine Vision

Roy C. Harrell
Daniel J. Cantliffe

Agricultural Engineering and Vegetable Crops Departments
University of Florida
Gainesville, Florida

I. INTRODUCTION

The analysis of bioreactor-based somatic embryogenesis requires the quantification of biological development. This involves the characterization of biological entities in terms of their numbers and distribution according to size and complexity of the cellular structures (Smith *et al.*, 1989). Machine vision is playing an expanding role in these types of measurements by providing the means to rapidly and objectively quantify plant tissue culture development. The efforts of Cazzulino *et al.* (1988) and Grand d'Esnon *et al.* (1988) have demonstrated the use of machine vision as an objective tool to quantify and qualify results of somatic embryogenesis research (see also Chapter 8, this volume). Cazzulino *et al.* (1988) utilized a machine vision system to estimate the number of carrot somatic embryos grown in suspension and the stage of development of each embryo. This information was used to verify a growth and development model of carrot somatic embryos grown in suspension. Grand d'Esnon *et al.* (1988) demonstrated the ability to recognize torpedo-shaped embryos of sweet potato and distinguish between embryogenic and nonembryogenic callus with machine vision.

Smith and Spomer (1987) and Smith *et al.* (1989) utilized machine vision to quantify the development of callus, shoot cultures, and whole plant microcultures grown on agar in contained culture vessels. This work was significant in that nondestructive analysis was achieved. In the case of Grand d'Esnon *et al.* (1988) and Cazzulino *et al.* (1988), destructive sampling techniques were required to prepare a sample suitable for analysis by the vision system. However, the nondestructive techniques employed were not suitable for the analysis of bioreactor suspensions since direct imaging of the stationary cultures was possible through the vessels. The noninvasive quantification of plant tissue development within a bioreactor has not yet been reported.

The impact of machine vision technology on the scale-up of somatic embryogenesis has been limited due to its destructive sampling requirement. Preparation of samples for analysis from a bioreactor typically requires drawing an aliquot of suspension with pipet. This sampling technique risks contaminating the suspension and is labor intensive. Both of these factors limit the number of samples which can be drawn and analyzed and, thus, limit the number of experiments that can be effectively quantified. Machine vision analysis in conjunction with an automatic, nondestructive sampling procedure would be capable of quantifying a suspension population without disrupting normal development patterns and with minimal, manual supervision. This measurement capability could improve the understanding of somatic embryogenesis by significantly increasing the number of bioreactor-based experiments being performed.

The objective of our work was to develop a machine vision measurement system which could noninvasively and automatically quantify the development characteristics of a heterogeneous plant tissue culture suspension grown in a bioreactor. Specifically, it was desired to periodically quantify the number and size distribution of calli of *Ipomoea batatas* Poir. cv. White Star (sweet potato) produced in a bioreactor without disrupting normal development. This chapter reports on the design, theory, and operation of the system developed to meet this goal.

II. MATERIALS AND METHODS

The calli for this study were initiated from shoot apices of *Ipomoea batatas* Poir. cv. White Star and proliferated on a solidified medium. The callus production (CP) medium contained the inorganic salts of Murashige and Skoog (1962), 500 μM *myo*-inositol, 5 μM thiamine-HCL, 10 μM nicotinic acid, 5 μM pyridoxine-HCL, 3% w/v sucrose, 10 μM 2,4-dichlorophenoxy acetic acid (2,4-D), and 1 μM bezylaminopurine (Chée and Cantliffe, 1988). Medium pH was adjusted to 5.8 using a 1 N sodium hydroxide solution before autoclaving at 121° C for 15 min. The medium was solidified by adding 0.7% w/v Phytagar (GIBCO Laboratories). Stock embryogenic suspension cultures were initiated from 8-week-old calli grown on solidified medium and maintained by subculture every 14 days into 500 ml Erlenmeyer flasks filled with liquid CP medium. Charge cultures for the bioreactor studies were obtained from these flasks.

Embryogenic calli (Color Plate 8A and 9A) were yellow, opaque, and firm with a nodulated surface, and nonembryogenic calli (Color Plate 9A) were white to brownish, translucent, friable, and somewhat uniform in structure (Chée and Cantliffe, 1989). When 2,4-D was omitted from the medium, embryo clusters formed on the embryogenic callus (Color Plate 8B). Embryogenic callus of 710–1000 μm in size (Color Plate 9A) was commonly used to examine callus growth. Embryos were readily produced from these callus clusters (Color Plate 9B) but were not used in this study.

III. SYSTEM DESCRIPTION AND OPERATION

The inspection station consisted of an airlift bioreactor (Kontes Cytolift) with a suspension transport system (Fig. 1) and a color machine vision system (Fig. 2). Analysis of the bioreactor's contents was accomplished by pumping the suspension from the reactor vessel through an imaging cell and back into the reactor. As the suspension passed through the imaging cell, images were obtained by the computer and analyzed to determine the number, size, shape, and color of aggregates within each image. The results from multiple images could be combined to estimate the population in suspension at the time of analysis and distribution of the population based on size and type of particles. The results from multiple analyses could be combined to estimate the time varying nature of these significant characteristics.

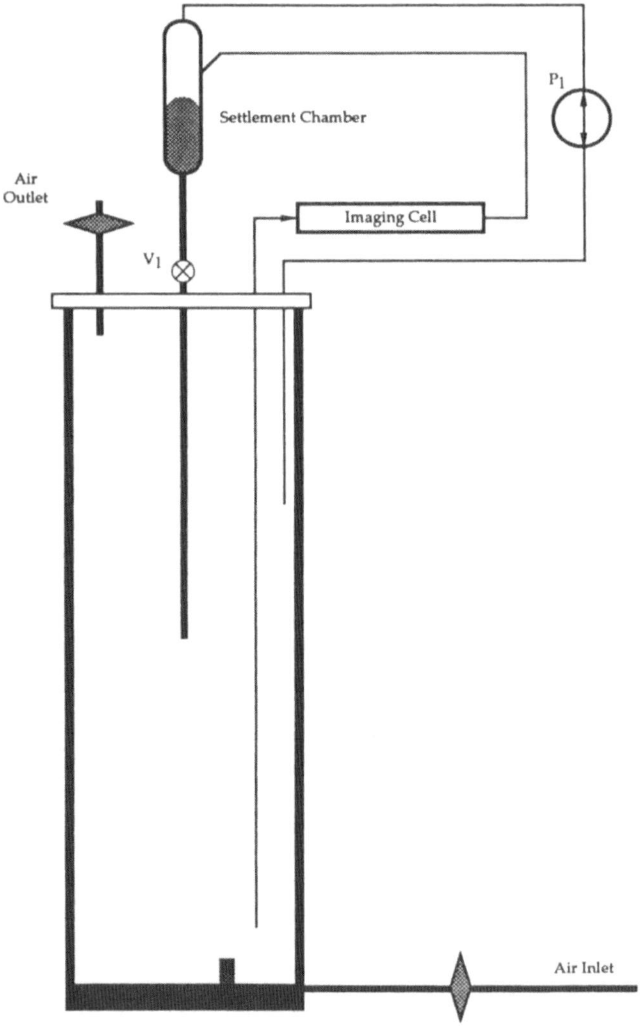

Fig. 1. The bioreactor and suspension handling system.

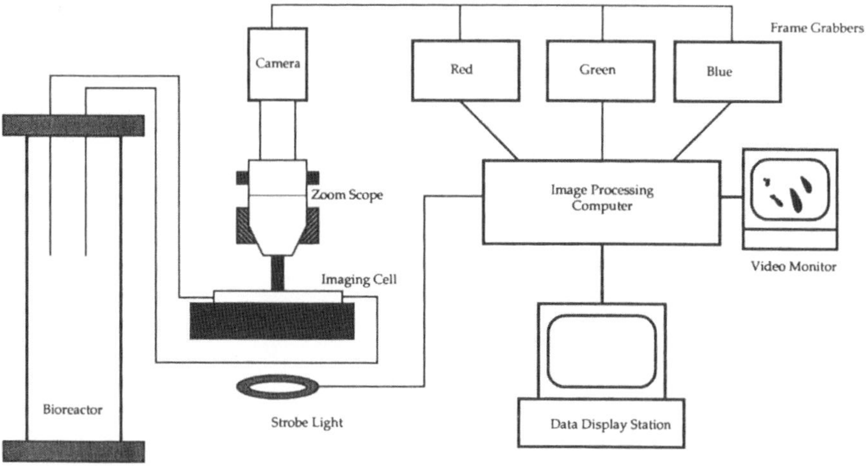

Fig. 2. Layout of inspection station components. Bioreactor details omitted.

The suspension transport system cycled the tissue culture suspension through the imaging cell and back to the reactor without damaging the aggregates. Suspension transport was accomplished with a peristaltic pump, a 3 mm wide × 3 mm deep × 50 mm long flow through imaging cell constructed from optical glass, and a settlement chamber. The settlement chamber separated medium and aggregates before the medium passed through the peristaltic pump. This was necessary to prevent damage to the aggregates by the pump. The reactor and suspension handling system were interconnected with silicon tubing. The assembled system (less pump) was autoclavable.

The machine vision system counted and quantified aggregate characteristics as the suspension passed through the imaging cell. The machine vision component included a stereoscopic zoom microscope (Nikon SMZ-2T), a color video camera (Pulnix TMC-54G), a 32-bit computer (Mizar 7120) equipped with three image acquisition cards (Datacube VVG128), and a custom stroboscopic illuminator. The three image acquisition cards were used to digitize the Red-Green-Blue (RGB) video signals generated by the camera. Each video signal was digitized with 5 bit resolution, which resulted in 32,768 unique colors distinguishable by the vision system. The flash of the stroboscopic illuminator was synchronized with image acquisition to produce nonblurred images of aggregates as they flowed through the imaging cell. The field of view of the vision system corresponded to a 2.7 mm (perpendicular to suspension flow) × 3.9 mm (parallel to suspension flow) area of the imaging cell.

The spatial resolution of image pixels was 28.6 μm (perpendicular to suspension flow) $\times$ 16.4 μm (parallel to suspension flow) which resulted in an active image array of 93 $\times$ 238 pixels.

IV. IMAGE ANALYSIS

Grand d'Esnon *et al.* (1988) used a gray scale vision system to distinguish between embryogenic and nonembryogenic callus by relating aggregate type to its mean transparency and size. The inability of the vision system to take advantage of the color difference between the embryogenic and nonembryogenic calli made this approach sensitive to variations in lighting and variations in density of calli within each type. The approach taken in our work was to distinguish the two calli based on color. This was accomplished by chromatic segmentation of an image into three classes: embryogenic callus, nonembryogenic callus, and background. The following discussion details the mechanics of the color segmentation technique.

Each unique color quantified by the vision system was represented as a three dimensional feature vector, **x**, as follows:

$$\mathbf{x} = [\mathrm{R\ G\ B}]^T$$

where vector elements quantified the intensities of the red, green, and blue primaries with values from 0 to 31. A pixel was assigned to the class embryogenic (w_e), nonembryogenic (w_n) or background (w_b) according to its color with the following relationship:

IF $P(w_e \mid \mathbf{x}) > P(w_n \mid \mathbf{x})$
    IF $P(w_e \mid \mathbf{x}) > P(w_b \mid \mathbf{x}))$
        THEN classify pixel as embryogenic
ELSE
    IF $P(w_n \mid \mathbf{x}) > P(w_b \mid \mathbf{x})$
        THEN classify pixel as nonembryogenic
    ELSE classify pixel as background (1)

where $\mathbf{x}$ = color of the pixel being classified,
$P(w_e \mid \mathbf{x})$ = the conditional probability that a color belongs to the embryogenic class,
$P(w_n \mid \mathbf{x})$ = the conditional probability that a color belongs to the nonembryogenic class, and

$P(w_b \mid \mathbf{x})$ = the conditional probability that a color belongs to the background class.

The conditional probabilities were determined with Bayes' Rule as follows (Snyder, 1985):

$$P(w_i \mid \mathbf{x}) = \frac{p(\mathbf{x} \mid w_i)P(w_i)}{[p(\mathbf{x} \mid w_e)P(w_e) + p(\mathbf{x} \mid w_n)P(w_n) + p(\mathbf{x} \mid w_b)P(w_b)]} \quad (2)$$

where $p(\mathbf{x} \mid w_i)$ = the conditional probability density function for class i (i = e for the embryogenic class, i = n for the nonembryogenic class, i = b for the background class); this represents the probability that a member of class w_i will have a feature vector **x** and

$P(w_i)$ = the *a priori* probability a feature vector belongs to class i.

It was assumed that probability densities could be approximated as Gaussian functions. Accordingly, the probability density functions were estimated with:

$$p(\mathbf{x} \mid w_i) = \frac{\exp\left[-\frac{1}{2}(\mathbf{x} - \mu_i)^T C_i^{-1}(\mathbf{x} - \mu_i)\right]}{(2\pi)^{d/2}\,|C_i|^{1/2}} \quad (3)$$

where μ_i = the mean vector of a training set, $\mathbf{X}_i$, containing N_i samples all known to belong to class i,

C_i = the covariance matrix of $\mathbf{X}_i$, and

$d = 3$ = the dimension of the feature space.

The mean vector of a training set was calculated with:

$$\mu_i = \frac{1}{N_i}\sum_{\mathbf{x}\in \mathbf{X}_i} \mathbf{x} \quad (4)$$

and the covariance matrix with:

$$C_i = \frac{1}{N_i}\sum_{\mathbf{x}\in \mathbf{X}_i} (\mathbf{x} - \mu_i)(\mathbf{x} - \mu_i)^T \quad (5)$$

This approach to pixel classification was implemented in the following manner. An interactive training session was used to construct the training sets $\mathbf{X}_e$, $\mathbf{X}_n$, and $\mathbf{X}_b$. This session allowed an operator to capture an image, position a 5 pixel × 5 pixel cursor anywhere in the image, and specify the class of the pixels in the cursor. For example, an image containing a nonembryogenic aggregate, an embryogenic aggregate, and background would be acquired. The operator would position the cursor until all cursor pixels were within the definable boundary of the embryogenic aggregate and assign them to the embryogenic class. The RGB values for the 25 cursor pixels would be included in the training set $\mathbf{X}_e$. The operator would reposition the cursor until it was within the boundary of the nonembryogenic aggregate and assign the new cursor pixels as nonembryogenic. The new RGB values would be included in the training set $\mathbf{X}_n$. The training session would continue until the operator was satisfied that sufficient data had been acquired to adequately describe the three classes. Data from multiple images could be incorporated into the three training sets.

After the training sets had been constructed, Eqs. (4) and (5) were solved to estimate the mean vectors and covariance matrices for the three classes. Next, Eq. (3) was solved three times for all of the 32,768 definable colors; once with i = e, once with i = n, and once for i = b. After the operator entered the *a priori* probabilities of the three classes, Eq. (2) was solved to determine the conditional probabilities for each color for each class. The denominator in Eq. (2) was not included in these calculations since it was constant for a given color. Each unique color was classified as embryogenic, nonembryogenic, or background, according to the logic in Eq. (1). The color classification information for all possible colors was stored in a $32 \times 32 \times 32 \times 1$ byte lookup table. This lookup table provided rapid access to color classification criteria by image processing routines used during on-line inspection of a bioreactor. Each color was represented in the lookup table by a byte that was set to 0 if classified as background, 1 if embryogenic, and 2 if nonembryogenic. On-line pixel classification was accomplished by retrieving a pixel's red, green, and blue intensity levels from the image planes and using these values as lookup table indices to the appropriate classification byte.

V. THE MEASUREMENT PROCESS

Under normal conditions, the bioreactor was operated with all particles in the reactor vessel and no particles in either the window loop or settlement chamber. Aggregates were uniformly suspended within the reactor with an airflow of approximately 0.1 liters/min pumped through a 0.2 μm filter. Valve V_1 was closed and pump P_1 off.

A measurement run would be initiated when it was desired to quantify the contents of the reactor. During a measurement run, the suspension in the reactor would be pumped through the imaging cell and into the settlement chamber. The suspension would be periodically imaged by the vision system as it passed through the imaging cell. Image analysis algorithms were developed to determine the number, size, and composition of aggregates captured in an image. Aggregate size was quantified according to its area, diameter, and perimeter. The composition of an aggregate was determined based on the percentage of its pixels, which were classified as embryogenic, nonembryogenic, and background. Images could be analyzed at 2 HZ.

Suspension transport through the imaging cell was accomplished by activating the peristaltic pump P_1 (see Fig. 1). Upon entry into the settlement chamber, cell aggregates would settle to the bottom and remain there during the course of the measurement run. The aggregate free medium was recycled to the bioreactor through P_1, maintaining a constant volume of medium in the reactor vessel. At the completion of a measurement run, the settlement chamber was flushed to return aggregates captured in the settlement chamber to the bioreactor. This was accomplished by temporarily shutting off air flow in the reactor, opening valve V_1, and reversing the direction of pump P_1.

The vision system could not directly measure the population in the reactor since several aggregates could pass through the imaging cell unobserved during the 0.5 sec period between image samples. Therefore, it was necessary to estimate the reactor population from the sampled data provided by the vision system. To do this, the relationship between the number of aggregates in the reactor at the start of a measurement run and the number observed by the vision system during the measurement run needed to be established. This relationship is developed below. It is assumed in this development that the reactor volume and suspension flow rate are known and constant during a measurement run. It is also assumed that aggregates in the reactor are uniformly suspended and that a constant sample period is used throughout a measurement run.

A population density balance on the reactor during a measurement run is expressed as:

Number of aggregates in − Number of aggregates out =
Change in number of aggregates

or

$$\rho_{\mathrm{i}}(t)q - \rho_{\mathrm{o}}(t)q = V_{\mathrm{r}}\frac{d\rho_{\mathrm{r}}(t)}{dt} \tag{6}$$

where ρ_i = Population density of incoming suspension, aggregates per milliliter
ρ_o = Population density of outgoing suspension, aggregates per milliliter
ρ_r = Population density of suspension in reactor, aggregates per milliliter
q = Volumetric flow of suspension, milliliters per minute, and
V_r = Volume of suspension in the reactor, milliliters.

Since the settlement chamber removes all aggregates from the suspension returning to the reactor ($\rho_i = 0$) and the aggregates in suspension are uniformly distributed ($\rho_r = \rho_o$), Eq. (6) reduces to the following:

$$\tau \frac{d\rho_r(t)}{dt} + \rho_r(t) = 0 \tag{7}$$

where $\tau = V_r/q$ and has units of minutes. Let $P_r(t)$ be the number of aggregates remaining in the reactor at some time t after the initiation of a measurement run. An expression for $P_r(t)$ is developed by substituting $\rho_r(t) = P_r(t)/V_r$ in Eq. (7) and solving with an initial condition of $P_r(0) = P_{ri}$ (the number of aggregates in the reactor at the start of a measurement run):

$$P_r(t) = P_{ri}e^{-t/\tau} \tag{8}$$

Let $P_w(t)$ be the number of aggregates which have passed through the imaging cell at time t. $P_w(t)$ can be expressed as the difference between P_{ri} and $P_r(t)$:

$$P_w(t) = P_{ri}(1 - e^{-t/\tau}) \tag{9}$$

Solving for P_{ri}:

$$P_{ri} = \frac{P_w(t)}{(1 - e^{-t/\tau})} \tag{10}$$

Equation (10) can be used to estimate the population in the reactor at some time t provided that q, V_r, and $P_w(t)$ are known. The reactor volume and suspension flow rate are constant during a measurement run and easily quantified. However, not all particles are observed by the vision system, and $P_w(t)$ must be estimated from the number observed. The ratio of observed particles to total particles is expressed as:

$$\frac{P_w(t)}{P_o(t)} = \frac{qT}{V_w} \tag{11}$$

where $P_o(t)$ = Total number of particles observed by the vision system at time t,
T = Vision sample interval, in minutes, and
V_w = Volume of imaging cell in field of view of vision system, in milliliters.

The numerator of this ratio is the total volume of suspension which passes through the imaging cell during a sample interval. The denominator is the volume of suspension analyzed by the vision system during a sample interval. Solving Eq. (11) for $P_w(t)$ and substituting in Eq. (10) results in:

$$P_{ri} = \frac{P_o(t)qT}{V_w(1 - e^{-t/\tau})} \qquad (12)$$

Equation (12) can be used to estimate the initial population of particles in the reactor based on the number of particles observed by the vision system.

VI. VERIFICATION OF THE INSPECTION SYSTEM

A series of tests were conducted to verify the ability of the bioreactor inspection station to quantify properties of a heterogeneous suspension. This was accomplished with a set of reference suspensions having known and constant properties. These suspensions consisted of small resin spheres (approximately 355 to 710 μm in diameter) suspended in water. To produce the reference suspensions, the resin spheres were sieved through 710 μm, 500 μm, and 355 μm nylon sieves. The >710 μm fraction and <355 μm fraction were discarded. The 500 μm and 355 μm fractions were characterized by manually weighing, counting, and sizing subsamples from each fraction (Tsukuda, 1989). Sizing of spheres was based on diameter as measured with a micrometer. A summary of sphere properties is given in Table I.

The first reference suspension was produced with 577 mg of the 355–500 μm fraction (approximately 10,000 particles) suspended in 500 ml of water in the reactor. Computer and manual measurements of the suspension were made. Manual measurements were performed by removing a 5 ml aliquot from the suspension, counting the particles in the aliquot,

TABLE I

Density and Size of Resin Spheres Used to Make Up Reference Suspensions

Suspension property	335–500 μm Fraction	500–710 μm Fraction
Density, particles/g	17,314	11,161
Density standard error, particles/g	832 (5%)	1,742 (16%)
Average diameter, μm	407	505
Diameter standard error, μm	79 (19%)	49 (10%)

and replacing the sample. Ten samples were analyzed in this fashion. Ten measurement runs were performed with the vision system. For each run, the reference suspension was pumped at 83.3 ml/min for 6 min. At the end of a measurement run, the particles contained in the settlement chamber were returned to the reactor in the manner described earlier. For the second reference suspension, an additional 578 mg of the 355–500 μm fraction were added to the reactor. Approximately 20,000 particles were in suspension for this test. Manual and computer analyses were repeated (10 runs each). For the third reference suspension, 1798 mg of the 500–710 μm fraction, or approximately 20,000 particles, were added to the reactor, yielding a total particle population of approximately 40,000. Manual and computer analyses were repeated on this suspension (5 manual samples and 10 computer runs).

Results from the verification tests are summarized in Table II and Fig. 3. The means and repeatability of the three measurement techniques for the three reference suspensions are shown in Table II. Means and repeatability for manual density estimates were obtained from data reported in Table I. The repeatability of computer analysis estimates was superior to either the mass-based estimates or the manual aliquot estimates. Population measurements for mass-based estimates and computer analyses were more consistent with each other than with measurements obtained by the manual aliquot technique.

TABLE II

Population Estimate Summaries

	Suspension No. 1	Suspension No. 2	Suspension No. 3
Mass-based population estimate	9,997 (±5%)	19,996 (±5%)	40,059 (±10%)
Manual aliquot population estimate	9,850 (±14%)	17,840 (±13%)	32,940 (±4%)
Computer estimate	11,169 (±6%)	21,484 (±6%)	39,100 (±4%)

Note: Percentages are repeatability of measurements as determined by ±1 standard deviation of measurement uncertainty.

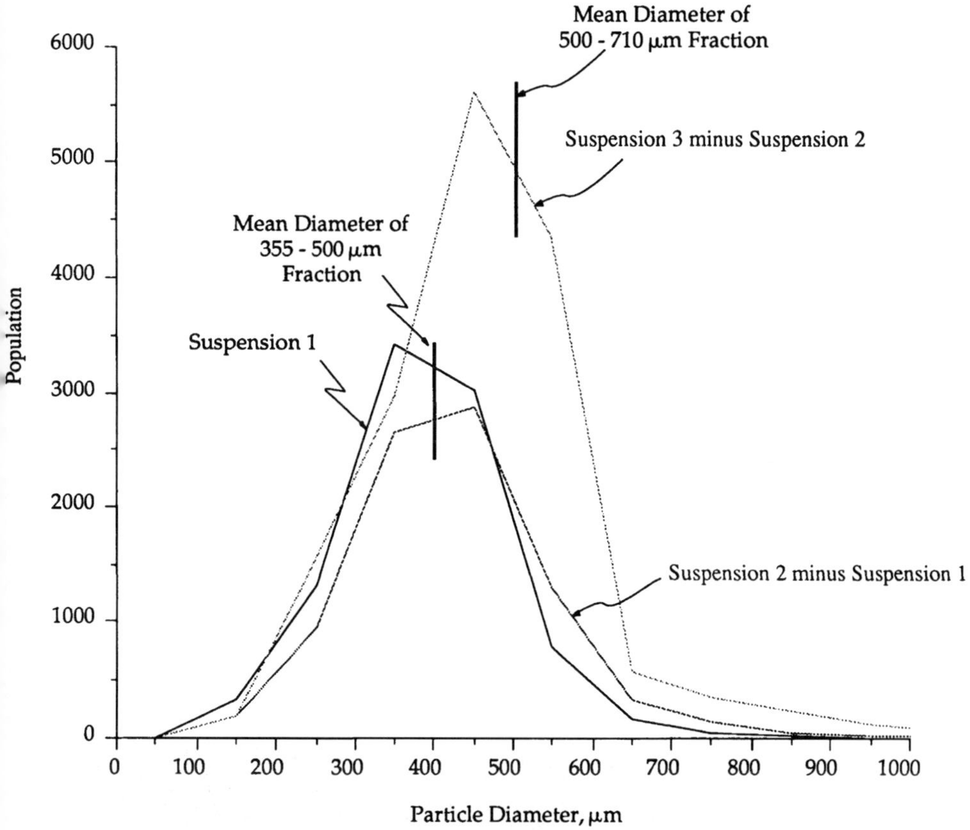

Fig. 3. Size distribution of the reference suspensions.

Diameter histograms of the sphere populations are shown in Fig. 3. The data plotted in Fig. 3 were obtained from the computer analyses of the suspensions. The horizontal axis is the particle diameter and the vertical axis is the number of particles observed at this diameter. Diameters were grouped into 100 μm resolution brackets. Three histograms are shown; suspension 1, suspension 2 – suspension 1, and suspension 3 – suspension 2. Suspension 1 histogram is the composite average of all particle size data obtained from the 10 computer runs on the first suspension. The suspension 2 – suspension 1 histogram is the average histogram from suspension 1 subtracted from the composite average histogram from suspension 2 and represents the histogram of the particles added to suspension 1 to obtain suspension 2. The suspension 3 – suspension 2 histogram is the average histogram from suspension 2 subtracted from the average histogram from suspension 3 and represents the histogram of the particles added to suspension 2 to create suspension 3. Histogram differences are plotted for the second and third suspensions since they reveal changes in the characteristics of a suspension and should be representative of the material added to the reactor. Histogram differences are important when attempting to quantify temporal changes in the characteristics of a population in suspension.

As expected, histograms for suspension 1 and suspension 2 − suspension 1 are very similar since the number of spheres used in suspension 1 was doubled to create suspension 2. Note that the peaks of these histograms correspond closely to the mean diameter of the particles used to create these suspensions (407 μm). Note also that the peak of the suspension 3 − suspension 2 histogram is approximately twice the height of the others and shifted by approximately 100 μm to the right. This was expected since the reactor population was doubled from suspension 2 to suspension 3 by the addition of approximately 20,000 spheres with an average diameter of 505 μm (see Table I).

VII. ON-LINE ANALYSIS

The objective of this test was to demonstrate the ability to noninvasively monitor the population characteristics of *Ipomoea batatas* Poir. cv. White Star suspended in a bioreactor. The reactor and suspension handling system were filled with 1 liter of liquid CP medium and charged with approximately 1 g of predominantly embryogenic aggregates which passed through a 1000 μm sieve and collected on a 710 μm sieve. Embryogenic aggregates used to charge the reactor were obtained from stock cultures suspended in 500 ml Erlenmeyer flasks with 80 ml of liquid CP medium for 2 weeks (Bieniek, 1990). After set-up, two or more runs were made on the suspension each day. For each run, the suspension was transported through the imaging cell at a flow rate of 83.3 ml/min for 30 min. After a measurement run, the aggregates collected in the settlement chamber were returned to the bioreactor in the manner previously described.

Population estimates for day 1, day 4, and day 5 for calli which were classified as having 50 percent or greater embryogenic material (as indicated by color) were 952, 1156, and 1298 aggregates, respectively. Diameter histograms for the 50 percent embryogenic population are shown in Fig. 4. The horizontal axis is aggregate diameter (micrometers) and the vertical axis is the number of aggregates within the size cate-

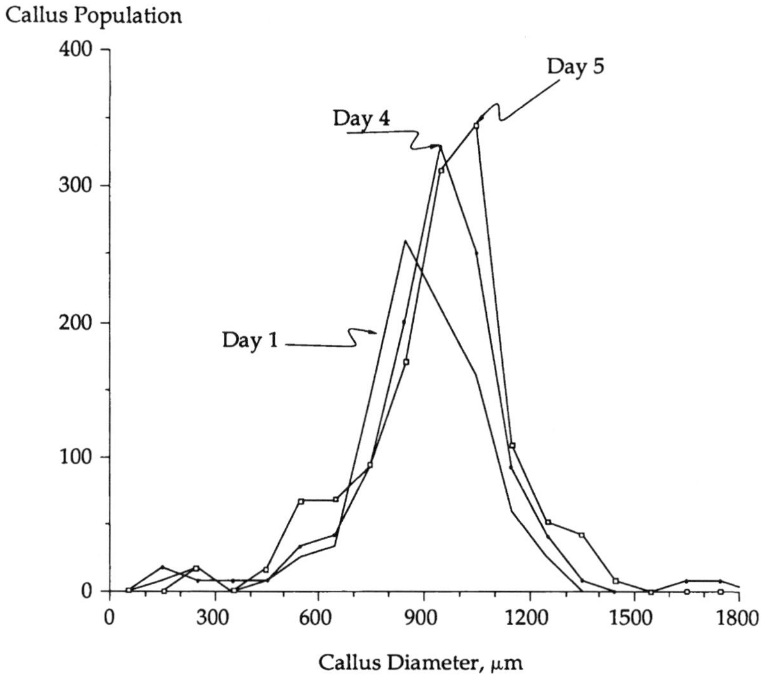

Fig. 4. Diameter histograms for calli in suspension that are classified as having 50% or greater embryogenic material. Day 1 population for this species was estimated at 952 aggregates, day 4 population estimated at 1156 aggregates, and day 5 population estimated at 1298 aggregates.

gory. A 100 μm diameter bracket was used. The data plotted in Fig. 4 were the averaged estimates for each size category for each day. An increase in the amount of embryogenic material in suspension is evident by the increase in mean diameter of the calli (as indicated by the peaks of the histograms shifting to the right) and an increase in the magnitudes of the peaks.

VIII. CONCLUSIONS

A machine vision–based measurement system was developed which could noninvasively quantify the population characteristics of a heterogeneous plant tissue culture suspension grown in a bioreactor. The vision system could estimate the total population of aggregates in suspension and produce size-based histograms of the population. In addition, aggregate types which differed in color could be distinguished by this system. Verification tests of the system demonstrated a worst case measurement repeatability of ±6%. This was superior to the measurement repeatabilities of aliquot-based estimates and mass-based estimates. The ability to estimate the size distribution of a heterogeneous suspension and changes in the size distribution characteristics of a suspension was demonstrated. The system was demonstrated by estimating the size distribution of a *Ipomoea batatas* Poir. cv. White Star culture suspended in an airlift reactor.

The measurement capabilities demonstrated by this system should be of value to the scale-up efforts of somatic embryogenesis. Noninvasive and automatic population analysis allows for the quantitative comparison of different growth strategies which would be impractical otherwise. The capabilities of the system could be easily expanded by incorporating embryo identification techniques as described in Grand d'Esnon *et al.* (1988), Cazzulino *et al.* (1988), and in Chapter 8, this volume. With the ability to distinguish between different types of embryos (e.g., globular, heart, torpedo, and aberrant stages), a population distribution of embryo types could be estimated. The time varying nature of this distribution could provide insights into the development morphology of somatic embryogenesis, leading to a better understanding of this complex biological phenomenon.

The vision capabilities demonstrated in this work are currently being incorporated into an embryo and callus harvester for an automated synthetic seed production system. The vision system will be used to classify the output of the reactor into one of the following categories: mature embryos, nonmature embryos, embryogenic callus, and nonembryogenic callus. The harvesting mechanism will be located immediately downstream from the imaging cell and used to route the different class particles to appropriate destinations. Particles which are identified as mature embryos by the vision system will be selectively removed by the harvester immediately after identification for post-harvest processing into a synthetic seed. Immature embryos and embryogenic calli will be recycled through the reactor. Nonembryogenic calli will be purged from the system by the harvester. Selective harvest of mature embryos could minimize the effects of nonsynchrony which commonly occur in somatic embryogenesis. Purging the nonviable material from a reactor could improve the efficiency of the overall process.

ACKNOWLEDGMENTS

The authors wish to acknowledge Roberto Munilla and Marie Bieniek for their assistance in the planning and setup of the bioreactor inspection system.

REFERENCES

Bieniek, M. (1990). Protocol for somatic embryogenesis of *Ipomoea batatas* Poir. Internal laboratory document. Synthetic Seed Laboratory, Vegetable Crops Department, IFAS, University of Florida, Gainesville. 32611. 8 pp.

Cazzulino, D. L., Pedersen, H., Chin, C. K., and Styer, D. (1988). Kinetics of carrot somatic embryo development in suspension culture. paper presented at the 1988 Annual Meeting of the AIChE, Washington, DC (Nov 27–Dec 2) 18 pp.

Chée, R., and Cantliffe, D. J. (1989). Composition of embryogenic suspension cultures of *Ipomoea batatas* Poir. and production of individualized embryos. *Plant Cell, Tissue Organ Cult.* **17,** 39–52.

Chée, R., and Cantliffe, D. J. (1988). Selective enhancement of *Ipomoea batatas* Poir. embryogenic and non-embryogenic callus growth and production of embryos in liquid culture. *Plant Cell, Tissue Organ Cult.* **15,** 149–159.

Grand d'Esnon, A., Chée, R., Harrell, R. C., and Cantliffe, D. J. (1988). Qualitative and Quantitative evaluation of liquid tissue culture by artificial vision. paper No. 88.395, presented at **AG ENG 88,** Agricultural Engineering International Conference, Paris, March 2–5, 12 pp.

Murashige, T., and Skoog, F. (1962). A revised media for rapid growth and bioassays with tobacco tissue culture. *Physiol. Plant* **15,** 473–497.

Smith, M. A. L., and Spomer, A. (1987). Direct quantification of in vitro cell growth through image analysis. *In Vitro Cell Devel. Biol.* **23,** 67–74.

Smith, M. A. L., Spomer, A., Meyer, M. J., McClelland, M. T. (1989). Non-invasive image analysis evaluation of growth during plant micropropagation. *Plant Cell, Tissue Organ Cult.* **19,** 91–102.

Snyder, W. E. (1985). "Industrial robots: computer interfacing and control." Prentice-Hall Inc. Englewood Cliffs, NJ.

Tsukuda, S. M. (1989). Performance quantification of an experimental bioreactor system. Report submitted to the Florida Foundation For Future Scientist, University of Florida, Gainesville. 32611. 30 pp.

CHAPTER **10**

Control of Contamination in Automated Plant Propagation

Alan C. Cassells

Department of Plant Science
University College
Cork, Ireland

I. INTRODUCTION

Attitudes to contamination, and to its control, vary widely in the micropropagation industry. Consequently, it is appropriate to preface this chapter with a brief discussion of perspectives on contamination. The micropropagator's view of contamination is influenced by several considerations. First, the degree of difficulty in establishing clean cultures varies widely among plant species. If one is working with a major crop, e.g., potato, the pathology of which is well known and for which commercial diagnostics are available, establishment of clean, that is, putative axenic, cultures may be an achievable goal, at least in geographical areas in which the crop has been historically studied. At the other extreme, there is the problem of introducing a less well-known plant from the tropics into culture (see Enjalric *et al.*, 1988). In the latter case, there may be a much greater diversity of potential contaminants associated with the tissue donor plant. A consequence of this diversity, and phytosanitary requirements for crops like potato, is that while most micropropagators seek to establish axenic cultures, others are more pragmatic and tolerate some contamination. Some rigorously monitor cultures and rogue out contamination, while others tolerate contamination as long as it does not have obvious economic consequences during production, even though only gross effects may be detectable. Among those who tolerate contamination, some seek to suppress it by manipulation of the medium, e.g., by manipulation of the hormone concentration (see Duhem *et al.*, 1988), while others may resort to the use of antibiotics in an attempt to suppress contaminants (see Hennerty *et al.*, 1988).

A shift in the structure of the industry from trade in *in vitro* cultures to in-house weaning of microplants and sale of liners or finished or partly finished plants (Jones, 1986; Cassells, 1991a) reduces the cost penalty associated with contaminated cultures in so far as the consumer would be unlikely to accept visibly contaminated cultures but is unlikely to test apparently healthy plants for endophytic organisms.

The above argument raises the question of how much contamination is, or should be, acceptable. The answer can only be made on the basis of adequate knowledge of the risks posed by contaminants which can be broadly categorized as follows (Cassells, 1991b):

1. Plant pathogenic microorganisms and other plant-associated microorganisms which may be latent through the micropropagation process or may be expressed at any stage *in vitro* or during crop development in the environment. Such organisms may pose a threat in new geographic regions.
2. Laboratory contaminants, including organisms associated with the operatives (Leifert *et al.*, 1989a), which may be expressed at any stage during culture or subsequent plant establishment and growing on. Such contaminants may carry drug resistance.

It is implicit that contaminants of either category may cause economic losses for the propagator during production or pose the threat of consequential loss claims when sold on. It is argued that the industry should observe high quality control standards, that is, seek to achieve proliferation based on axenic cultures and adequately monitor production for contaminants which should be carefully disposed of (Cassells, 1991a,b,c).

Automation of micropropagation implies a move to larger culture containers or vessels. This inevitably increases the potential problems associated with contamination, as the larger and more open the system and the more agitation, the greater the potential for inoculum dispersal within. Those involved in studies on plant cell fermentations are fully aware of the downtime due to contamination. Animal cell culturists incorporate wide-spectrum antibiotics into their media as standard practice (Perlman, 1979) and this practice may be tempting to micropropagators using parallel systems but poses the associated risk of the occurrence of drug resistant strains which in turn may threaten production and cause illness in the work force.

In this review, the problems of contamination in automation will be considered in the overall context of quality control standards for the industry. Automation has as a prerequisite a manual start-up phase. Contamination can be considered as being carried over into culture in or on the explant; if axenic cultures are not obtained, such contaminants may be transmitted at subculturing. When "clean" cultures have been obtained, they can be used to seed the automated process, and here a new source of contamination arises, namely, laboratory contamination which includes contaminants derived from the individuals involved in the process. The framework in which the problem of contamination in micropropagation will be addressed is shown in Fig. 1. Aspects of the problem have been also discussed elsewhere (Cassells, 1986, 1988, 1991a,b,c).

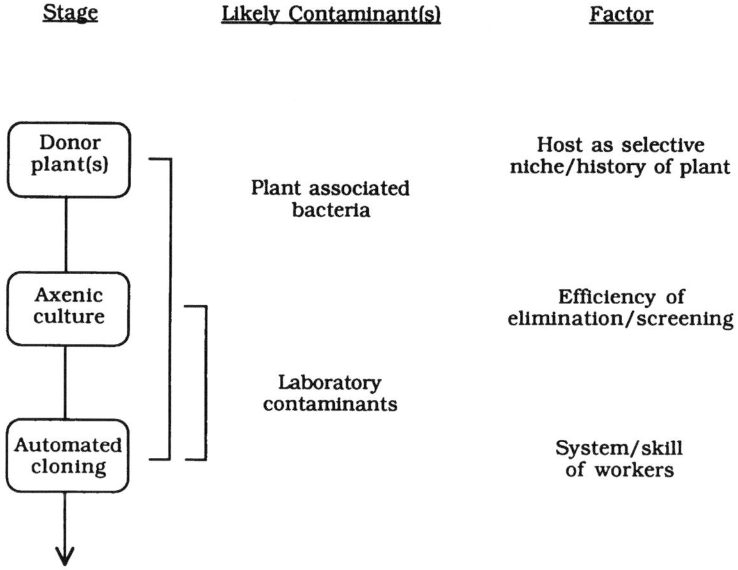

Fig. 1. Stages in automated micropropagation at which contamination may be encountered and factors influencing the contamination.

II. ORIGIN AND NATURE OF CONTAMINATION IN MICROPROPAGATION

The difficulty and implied cost of identifying contaminants generally deters micropropagators pursuing this aspect of the problem. The exception is specific known diseases of the crop which the market requires be eliminated. However, most micropropagators would distinguish between contaminants introduced on or in the explant and "laboratory" contamination. The latter, it is now appreciated, may include both heat-resistant species, e.g., *Bacillus* spp., and organisms derived from the atmosphere and human operative (Leifert *et al.*, 1989a) (Table I). The latter may potentially include human pathogenic species of bacteria and fungi and antibiotic-resistant isolates.

Here a distinction is made between cultivable and noncultivable organisms. The former have the potential to overrun the cultures and many proliferate in the tissues causing damage, while the effects of the latter on *in vitro* cultures have not been extensively studied (see Long *et al.*, 1988; Leifert *et al.*, 1989b) and may include pathogens or potential pathogens of the progeny plants.

TABLE I
Indicator Species Isolated from Contaminated Tissue Cultures[a]

Organism(s)	Indication
Klebsiella oxytoca and microorganisms in Table II	Failure to establish axenic culture: contaminated explants.
Bacillus pumilus, B. circulans, and other *Bacillus* species	Inadequate sterilization of media and/or equipment.
Lactobacillus acidophilus	Contamination originates from personnel.
Micrococcus kristinae	Contamination originates from personnel.
Pseudomonas aeruginosa, P. paucimobilis, P. cepacia, and other representatives of the normal microbial flora of the human body (see Jawetz *et al.*, 1989)	Contamination originates from personnel.
Staphylococcus epidermidis	Contamination originates from personnel.

[a]Based on Leifert *et al.*, 1989a.

TABLE II
Common Plant-Associated Bacterial Genera[a]

Genus	Gram stain
Agrobacterium	Negative
Rhizobium	Negative
Erwinia	Negative
Pseudomonas	
Xanthomonas	Negative
Corynebacterium[b]	Positive

[a]Based on Chatterjee and Vidaver, 1986.
[b]Many historical members of this genus have been transferred to other genera including *Rhodococcus, Nocardia, Curtobacterium,* and *Clavibacter.*

A. Contamination Arising from Plant-Associated Organisms

The noncultivable and intractable organisms included in the category of plant-associated organisms are the viruses and viroids, mollicutes, phloem, and xylem-limited bacteria. The mollicutes, phloem, and xylem-limited bacteria are potentially cultivable; however, they are fastidious and consequently do not generally pose the threat that they will proliferate on the culture medium and overrun the cultures.

Cultivable bacteria and yeasts, on the other hand, pose a serious risk of contamination, as most readily proliferate in the explant tissues and can invade the medium, depending on its composition, to varying degrees. The principal genera of plant-associated bacteria are given in Table II (see Cassells, 1991b,c).

B. Contamination Arising in the Laboratory

Cultivable organisms which may recontaminate cultures in the laboratory include common atmospheric or water-borne bacteria and yeasts. The detection of heat-resistant microorganisms such as *Bacillus* may indicate incomplete autoclaving of media and glassware (Table I), while the presence of yeasts may indicate a contaminated air supply. In a benchmark paper on this topic, Leifert *et al.* (1989a) reported the presence of bacteria associated with human tissues in cultures, progressively increasing with subculture.

Identification of contaminants merits serious cost–benefit analysis, as correct identification may help in locating the source of the contamination and thus facilitate control (see Table I).

III. IDENTIFICATION OF CONTAMINANTS

A. Plant-Associated Microorganisms

Identification to some level, be it genus or species, is essential before control strategies can be implemented. This is especially the case if antibiotics are to be used (see Falkiner, 1988; 1990).

Screening can be carried out effectively at two levels, namely, for known pathogens of the crop and for cultivable organisms (Table III). For the former ELISA and DNA probes may be commercially available or test plants known. For the latter, yeasts and most plant-associated bacteria are aerobes and will grow on a range of simple media (Table IV; Leifert *et al.*, 1989a; Leifert *et al.*, 1990).

Extreme difficulty, however, can be encountered in detecting uncharacterized noncultivable or fastidious microorganisms. The possibility of

TABLE III
Screening for Pathogens and Contaminating Organisms in Micropropagation[a]

Stage	Action
0	Visually examine potential donor plants for disease symptoms, screen for cultivable bacteria, use streak plating to separate isolates, use serial dilution to determine contamination. Use specific tests for known pathogens of the crop. Reject heavily contaminated or diseased individuals if practical.
I	Visually examine cultures for contamination, including "halo" formation. Reject all contaminated cultures. Screen remaining cultures for cultivable organisms and carry out specific tests for known pathogens of the crop. Reject all infected cultures (treat donor plants and repeat meristem culture, see Cassells, 1986; Cassells *et al.*, 1988).
II	Monitor production: sample production, examine visually and carry out tests for cultivable organisms. If contamination is detected, screen upstream and downstream production, reject as appropriate.
III	As for Stage II.
IV	Established progeny should also be sampled and monitored principally for known diseases of the crop which may have been below the level of detection *in vitro*.

[a]Stages based on Debergh and Maene (1981) (from Cassells, 1991b).

TABLE IV
Methods Used in the Detection and Identification of Microorganisms in Tissue Culture[a]

Test	Application	Reference
Nonspecific tests		
Culture indexing	Cultivable bacteria	Schaad, 1980 Lelliott and Stead, 1987
DNA/RNA fluorochrome staining, e.g., DAPI	*In situ* staining of mycoplasmas and related prokaryotes	Möllers and Sarkar, 1989
Leaf dip electron microscopy	Rod morphology viruses, high titer viruses	Noordham, 1973 Hill, 1984
Gel electrophoresis	Viroids	Morris and Smith, 1977
Specific tests		
Enzyme linked immunosorbent assay (ELISA)	Viruses and bacteria	Clark and Adams, 1977 Hill, 1984
Protein A latex-linked antiserum test (PALIAS)	Viruses and bacteria	Querfurth and Paul, 1979 Hill, 1984
Immunosorbent electron microscopy (ISEM)	Viruses	Derrick, 1973 Hill, 1984
DNA probes	All organisms	Owens and Diener, 1981 Scharf *et al.*, 1986
Rapid diagnostic kits	Bacteria	Mergaerts *et al.*, 1984 Collins and Lyne, 1985 Lelliott and Stead, 1987
Fatty acid profiling	Bacteria	Stead, 1988

[a] From Cassells, 1990b.

the presence of such organisms challenges the concept of axenic cultures which at best may be termed putative axenic cultures.

B. Laboratory Contaminants

Assuming that plant-associated microorganisms have been eliminated in Stage 0 (Debergh and Maene, 1981) or by Stage I (Murashige, 1974), then the spectrum of contaminants will change to reflect laboratory contamination, and further, assuming that only cultivable organisms will be involved, then screening moves into the area of clinical microbiology, where commercial diagnostic kits may be applicable (Table V). Detection of contaminants may not be timely enough to take remedial action, but if combined with monitoring, preventive measures may be developed which are applicable in future production.

TABLE V
Characterization of Bacterial Contaminants from Plant Tissue Cultures[a]

Preliminary tests	Primary groupings	Characterization[b]
Gram stain Shape Mobility Catalase test Oxidase test Heat stability Oxidation/fermentation	Gram-negative rods (Enterobacteriaceae)	API 20 E and API 50 CHE strips
	Nonacidifying Gram-negative rods and cocci	API 20 NE
	Catalase-negative, Gram-positive rods (*Lactobacillus* spp.)	API 20 B and API 50 CHL
	Heat-resistant, Gram-positive rods (*Bacillus* spp.)	API 20 E and API 50 CHB
	Gram-positive cocci	API Staph tests

[a]Leifert *et al.*, 1989a.
[b]90% of the bacteria keyed out using API computer identification software; however, these results should not be taken as representative, as the Stage 0 and Stage I procedures used appear to have been particularly efficacious.

In general, it is unlikely that the opportunity will exist for reinfection by noncultivable or fastidious organisms under laboratory conditions. Possible exceptions are certain viruses which are mechanically transmissible (Hill, 1984) or vector-transmitted agents where cultures become contaminated with infectious vectors (Blake, 1988).

IV. CONTROL OF CONTAMINATION

A. Establishment of Clean Cultures

1. Via Stage 0 and Meristem Culture Procedures

As previously emphasized, ambiguity exists as to the status of cultures which may be attained. The norm to be aimed for may be cultures free of specified diseases of the crop and free of cultivable microorganisms (Table VI).

In many cases meristem culture *per se* may give clean cultures (Cassells, 1986). However, in the case of invasive viruses and persistent cultivable organisms donor plant treatment may be a prerequisite (Stage 0, of

Debergh and Maene, 1981). In the latter approach, the donor plant is grown under conditions of low fertilization which hardens cell walls and may restrict bacterial colonization and with bottom watering to reduce inoculum dispersal. Treatment of the donor plant with antibiotics may be used at this stage to reduce bacterial titer and invasiveness (Cassells *et al.*, 1988).

Attempts to eliminate microorganisms in Stage 0 or via meristem culture in Stage I are only as reliable as the methods used to screen the *in vitro* material for the presence of contaminants (Tables III and IV). In our experience limited multiplication in Stage II and elongation of shoots as in Stage III are a prerequisite for screening. Normally intercalary nodes would be subcultured individually and upper and lower nodes destructively screened. Low titer bacteria may be inhibited in sap extracts when a tissue homogenate is plated directly (Cassells, 1991b,c). In this case, the explant may be subjected to enrichment by placing the intact tissue on bacteriological media to facilitate the buildup of bacterial titer in the explant with subsequent expression on the medium.

A suggested alternative to sampling the cultures is to culture the explants on media enriched with bacterial nutrients to encourage expression (Kunneman and Faaij-Groenen, 1988). The high salt content of the basal plant tissue culture medium may inhibit the growth and expression of some plant-associated bacteria (Cassells *et al.*, 1988).

2. The Use of Antibiotics in *in Vitro* Cultures

Effective use of antibiotics in chemotherapy depends on (1) accurate identification of the target organism which allows selection of suitable antibiotics, (2) determination of the minimum inhibitory concentration of the antibiotic and, in this case, confirmation that the plant tissue cul-

TABLE VI
Health Certification of *In Vitro* Cultures[a]

Certification	Test result[b]
Class 1	Free of all cultivable microorganisms, free of all major diseases of the crop.
Class 2	Free of all cultivable microorganisms, free of specified diseases of the crop.
Class 3	Free of cultivable bacteria.
Class 4	Unscreened.

[a]From Cassells, 1991c.
[b]Tests must be based on adequate sampling of production. Screening for cultivable microorganisms as in, e.g., Lelliott and Stead, 1987; specific tests to be based on ELISA or similar accepted tests.

ture medium does not affect the action of the antibiotic (Falkiner, 1988), and (3) confirmation that the antibiotic does not cause plant tissue toxicity or mutagenesis (Moutschen, 1985).

Stage (1) can be short-circuited if one uses an appropriate broad-spectrum antibiotic based on the Gram-stain and other relatively easily determined characteristics of the bacterium (see Tables V and VII),

TABLE VII

Evaluation of Antibiotics for Use in Plant Tissue Culture[a]

Antibiotic groups	Observation
Inhibitors of cell wall synthesis (1) Bactracin (2) β-lactams (3) Glycopeptides	Internal pressure in Gram-positive bacteria is 3–5 times higher than in Gram-negative. L-forms stable in hypertonic environment ≥20% sucrose, i.e., may survive on plant tissue culture media. (1) Mainly bacteriocidal for Gram-positive organisms, including those resistant to β-lactams. (2) Bacteriocidal, mainly Gram-positive organisms, 10–100× concentration may be required for Gram-negative. Resistance common. Allergenic. (3) Bacteriocidal for Gram-positives only, expensive but resistance uncommon.
Inhibitors of cell membrane function (1) Polymixins	Bacteriocidal for Gram-negatives, including *Pseudomonas*. Resistance rare.
Inhibitors of protein synthesis (1) Chloramphenicol (2) Aminoglycosides (3) Macrolides and Lincosamides (4) Tetracyclines	Usually reversible, toxicity follows long exposure. (1) Bacteriostatic, spectrum as tetracyclines. Cheap but phytotoxic. (2) Bacteriocidal, broad spectrum but may be phytotoxic, more active at alkaline pH's, used most widely against Gram-negative bacteria, clinical resistance known, can be used with penicillins. (3) Bacteriostatic for Gram-positive bacteria. Resistance frequent. Interference may occur between drugs of this group. (4) Bacteriostatic, to both susceptible Gram-positive and Gram-negative bacteria. Reported to stimulate yeasts, phytotoxicity reported.
Inhibitors of nucleic acid synthesis (1) Quinolones (2) Rifampicin (4) Trimethoprim	(1) Bacteriocidal, mainly Gram-positive organisms, 10–110×, concentration may be required for Gram-negative. Resistance common. Allergenic. (2) Bacteriocidal, some Gram-positive and Gram-negative bacteria. Resistance frequent. (3, 4) Bacteriostatic for some Gram-positive and Gram-negative bacteria. Marked enhancement if used together. Sulphonamides active at alkaline pH's. Resistance known. Can be allergenic.

[a]Jawetz *et al.*, 1989; Falkiner, 1990.

having ensured by spread plating that only one contaminant is involved. Should more than one be detected, it may be necessary to use a mixture of antibiotics, and here compatibility must be confirmed (Jawetz *et al.*, 1989; Falkiner, 1988; 1990). A major deterrent to the use of antibiotics is the fear that they may be mutagenic. The narrower the genetic base to production, the greater the risk (Cassells, 1991a).

Table VII lists antibiotics in current use and evaluates their potential in plant tissue culture. It should be appreciated that antibiotics have been developed for use in mammalian tissues and hence may be toxic to plants and that prolonged exposure, as in the long cycle of plant tissue culture, may result in resistance emerging or breakdown in bacteriostasis due to metabolism or instability of the antibiotic (Jawetz *et al.*, 1989). It is also obvious from the literature that in many cases plant tissue culturists have misused antibiotics by use of arbitrary doses, incompatible mixtures, etc., without any attempt to identify the contaminant(s) (see Falkiner, 1988, 1990). For a general discussion of this aspect see Jawetz *et al.* (1989).

V. MONITORING AND CONTROL OF CONTAMINATION DURING AUTOMATED PRODUCTION

A. Preparation for Automation

Regardless of the system of automation, culture inoculum for the process will have to pass through Stages 0, I, and possibly II to initiate the automated process with the problems and risks discussed in Section III. If the preparation of inoculum has been rigorous, then arguably the main contamination risk is posed by laboratory contaminants, which, if the previous reasoning is accepted, will include the possibility of exotic microorganisms arising from the atmosphere and production workers (Table I).

B. Automation

The design of the system greatly influences the problems associated with the control of contamination. The main design features are the ease with which the vessels can be sterilized and loaded with culture inoculum, the volume of the unit, and the opportunity for contaminant inoculum to be dispersed internally.

Two basic model micropropagation systems are represented in Fig. 2.

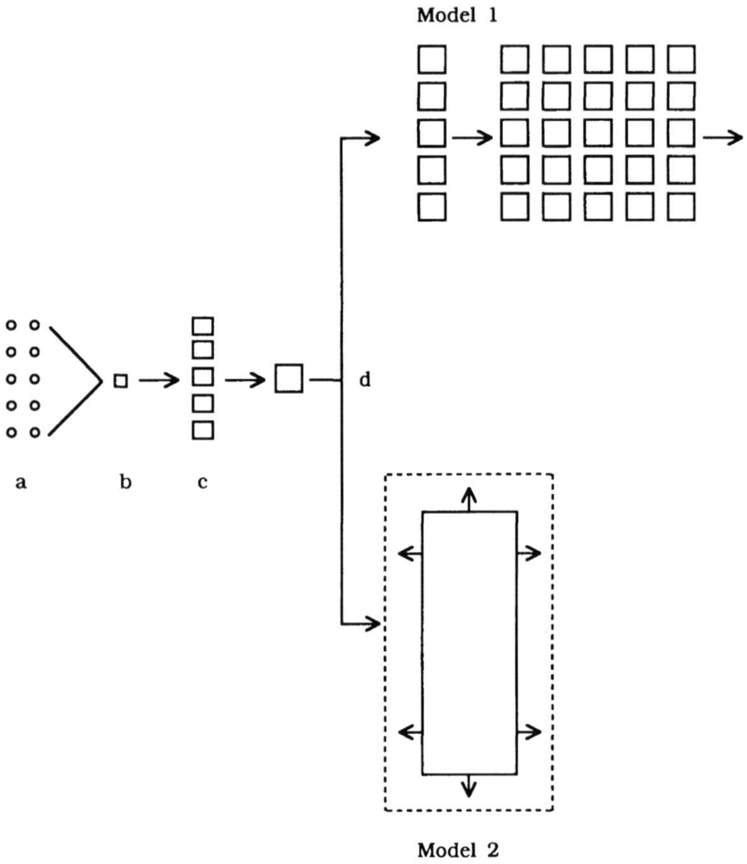

Fig. 2. Alternative strategies for automation. In both models steps (a), (b), and (c) are common. Step (a) involves the introduction of the plant into culture from an adequate genetic base (Stage I). Step (b) involves limited multiplication for contaminant screening and bulking of seed inoculum (Stage II). Step (c) is the manual transfer of axenic biomass to the standard container for an automated containerized process. In Model 2 step (c) may be omitted, the open reactor is seeded with axenic material from (b) or (c). Production in Model 1 takes place in sealed units which restrict contaminant movement. In Model 2 the system is open and no internal restriction of contaminants is possible.

In both, individual cultures are established from a broad genetic base (Cassells, 1991a). These cultures are passed to Stage II, where they are rigorously screened (Cassells, 1991b,c) and, if clean, are used as culture inoculum. Here the models diverge. In Model 1 the putative axenic material is manually transferred to production scale containers, containers are seeded, and production is then automated, but container size is fixed. In Model 2, a large chamber or vessel (e.g., a bioreactor) is seeded, and large-scale batch production results. In these two extremes, namely, the use of small containers versus a large open vessel, strategies for contamination control would differ. In the former, where production is compartmentalized, many of the protocols for contamination monitoring in conventional micropropagation can be employed (Table III; Cassells, 1991b,c). In Model 2 or any internally open system, contamination, once introduced, would be unrestricted. The contrast between the models is the distinction between restricted contaminant expression on the one hand and essentially batch fermentation of the contaminant on the other. Given that the open system is based on liquid culture, the medium could be circulated and filtered and/or monitored for contamination at short intervals after start-up for drug sensitivity testing and introduction of antibiotics as appropriate. Alternatively, broad-spectrum antibiotics and fungicides could be incorporated in the medium as a prophylactic measure. The latter strategy parallels the approach of animal cell culturists (Perlman, 1979).

There are clear risks associated with the prophylactic use of antibiotics in plant tissue culture which have been discussed by Falkiner (1990).

VI. CONCLUSION

The market for any product depends on quality and price. In the case of the micropropagation industry, reduction in production costs will increase profitability in existing markets and, if the cost reduction is significant, may lead to market expansion (Sluis and Walker, 1985). Automation of the division of organized tissues is one strategy aimed at reducing production costs (see Levin and Vasil, 1989). To realize the full economic benefits of this advance, automated transfer to the environment must also be achieved. Even then, however, the product has limited shelf life. Micropropagation may only become competitive with seed when automated somatic embryogenesis and artificial seed production is achieved (Redenbaugh, 1986; see also Chapters 3 and 8, this volume). While not explicitly covered here, contamination problems in large-scale plant cell culture and *in vitro* somatic embryogenesis are similar to those described for the large open vessel system described.

In the move to automation, contamination and deviation-from-type represent important aspects of quality control, and it should be borne in mind that control of the former may affect the latter, namely, when the control measures adversely affect genetic stability.

A general principle of contamination control is avoidance. In any form of micropropagation, primary contamination arises from contaminated explants, while secondary contamination arises from laboratory sources. Awareness of the sources and rigorous staff training are important elements in contamination control. It is argued here that the implementation of a control strategy depends on production of axenic cultures and careful monitoring of the preautomation stage. Contamination risks during automated production then depend largely on the system employed, large open systems posing particularly high risks, as monitoring may be frustrated and remedial action difficult. Risk avoidance in open systems may include the prophylactic use of broad-spectrum antibiotics and fungicides, but these can break down for the reasons discussed and they represent a recurrent cost.

Transcending the prophylactic use of antibiotics is the issue of health and safety. The use and abuse of antibiotics in animal husbandry has been widely discussed (see Falkiner, 1990). The micropropagation industry is faced with similar issues in a more environmental risk-conscious society. Concern for workers in the horticulture industry in some countries where antibiotics and pesticides are abused has aroused public concern. Similar abuse in micropropagation, where drug-resistant potential human pathogens may be transmitted, is a significant threat to public perception of the industry and its product.

Studies on the use of antibiotics in plant tissue cultures are increasing. Only detailed investigation will resolve the issues, and it is hoped that commercial companies will freely disclose results to enable an accurate costing to be made of losses due to contamination, which in turn will enable investigators to carry out a cost–benefit analysis of preventive strategies.

ACKNOWLEDGMENTS

The author is grateful to Professor W. Waites and co-workers and to Dr. F. Falkiner for permission to cite material in press.

REFERENCES

Blake, J. (1988). Mites and thrips as bacterial and fungal vectors between tissue cultures. *Acta Hortic.* **225,** 163–166.

Cassells, A. C. (1986). Production of healthy plants. *In* "Micropropagation in Horticulture:

Practice and Commercial Problems" (P. G. Alderson and W. M. Dullforce, eds.), pp. 53–70. Institute of Horticulture, London.
Cassells, A. C. (1988). Symposium on Bacterial and Bacteria-Like Contaminants of Plant Tissue Cultures. *Acta Hortic.* **225,** 1–225.
Cassells, A. C., Harmey, M. A., Carney, B. F., McCarthy, E., and McHugh, A. (1988). Problems posed by cultivable bacterial endophytes in the establishment of axenic cultures of *Pelargonium* x *domesticum:* the use of *Xanthomonas pelargonii*-specific ELISA, DNA probes and culture indexing in the screening of antibiotic treated and untreated donor plants. *Acta Hortic.* **225,** 153–162.
Cassells, A. C. (1991a). Setting up a commercial micropropagation laboratory. *In* "Biotechnology in Agriculture and Forestry" (Y. P. S. Bajaj, ed.). Springer-Verlag, Berlin. (in press).
Cassells, A. C. (1991b). Screening for pathogens and contaminating micro-organisms in micropropagation. *In* "Techniques for Rapid Detection and Diagnosis in Plant Pathology" (J. M. Duncan and C. Torrance, eds.). Blackwell, Oxford. (in press).
Cassells, A. C. (1991c). Problems in tissue culture: culture contamination. *In* "Micropropagation of Horticultural Crops" (R. H. Zimmerman and P. C. Debergh, eds.). pp. 31–44. Kluwer, Dordrecht.
Chatterjee, A. K., and Vidaver, A. K. (1986). "Genetics of pathogenicity factors: application to phytopathogenic bacteria." Advances in Plant Pathology, Vol. 4, pp. 1–35. Academic Press, New York.
Clark, M. F., and Adams, A. N. (1979). Characteristics of the microplate method of enzyme-linked immunosorbent assay for the detection of plant viruses. *Journal Gen. Virol.* **35,** 475–483.
Collins, C. H., and Lyne, P. M. (1985). "Microbiological Methods," (5th edn.), pp. 102–113. Butterworths, London.
Debergh, P. C., and Maene, L. J. (1981). A scheme for commercial propagation of ornamental plants by tissue culture. *Scientia Hortic.* **14,** 335–345.
Derrick, K. S. (1973). Quantitative assay for plant viruses using geologically specific electron microscopy. *Virology* **56,** 652–653.
Duhem, K., Le Mercier, N., and Boxus, Ph. (1988). Difficulties in the establishment of axenic *in vitro* cultures of field collected coffee and cacao germplasm. *Acta Hortic.* **225,** 67–75.
Enjalric, F., Carron, M. P., and Lardet, L. (1988). Contamination of primary cultures in tropical areas: the case of *Hevea brasiliensis. Acta Hortic.* **225,** 57–65.
Falkiner, F. R. (1988). Strategy for the selection of antibiotics for use against common bacterial pathogens and endophytes of plants. *Acta Hortic.* **225,** 53–56.
Falkiner, F. R. (1990). The criteria for choosing an antibiotic for control of bacteria in plant tissue cultures. *IAPTC Newsletter* **60,** 13–23.
Hennerty, M. J., Upton, M. E., Furlong, P. A., Harris, D. P., James, D. J., and Eaton, R. A. (1988). Microbial contamination of *in vitro* cultures of apple rootstocks M26 and M9. *Acta Hortic.* **225,** 129–137.
Hill, S. A. (1984). "Methods in Plant Virology." pp. 1–176. Blackwell, Oxford.
Jawetz, E., Melnick, J. L., Adelberg, E. A., Brooks, G. F., Butel, J. S., and Ornston, L. N. (1989). Medical Microbiology (18th edn.), pp. 143–172. Appleton and Lange, Norwalk.
Jones, J. B. (1986). Determining markets and market potential of horticultural crops. *In* "Tissue Culture as a Plant Production System for Horticultural Crops" (R. H. Zimmerman, R. J. Griesbach, F. A. Hammerschlag, and R. H. Lawson, eds.), pp. 175–182. Martinus Nijhoff, Dordrecht, The Netherlands.
Kunnemann, B. P. A. M., and Faaij-Groenen, G. P. M. (1988). Elimination of bacterial contaminants: a matter of detection and transplanting procedures. *Acta Hortic.* **225,** 183–188.

Leifert, C., Waites, W. M., and Nicholas, J. R. (1989a). Bacterial contamination of micropropagated plant cultures. *J. Appl. Bacteriol.* **67,** 353–361.

Leifert, C., Waites, W. M., Camotta, H., and Nicholas, J. R. (1989b). *Lactobacillus plantarum:* a deleterious contaminant of plant tissue cultures. *J. Appl. Bacteriol.* **67,** 363–370.

Leifert, C., Waites, W. M., Nicholas, J. R., and Keetley, J. W. (1990). Yeast contaminants of micropropagated plant cultures. *J. Appl. Bacteriol.* **69,** 471–476.

Lelliott, R. A., and Stead, D. E. (1987). Methods for the Diagnosis of Bacterial Diseases of Plants. pp. 1–216. Blackwell, Oxford.

Levin, R., and Vasil, I. K. (1989). Progress in reducing the cost of micropropagation. *IAPTC Newsletter* **59,** 2–12.

Long, R. D., Curtin, T. F., and Cassells, A. C. (1988). An investigation of the effects of bacterial contaminants on potato nodal cultures. *Acta Hortic.* **225,** 83–91.

Mergaerts, J., Verdonch, L., Kersters, K., Swings, J., Boeufgras, M. J., and De Ley, J. (1984). Numerical taxonomy of *Erwinia* species using API systems. *J. Gen. Microbiol.* **130,** 1893–1910.

Möllers, C., and Sarkar, S. (1989). Regeneration of healthy plants from *Catharanthus roseus* infected with mycoplasma-like organisms through callus culture. *Plant Sci.* **60,** 83–89.

Morris, T. J., and Smith, E. M. (1977). Potato spindle tuber viroid disease: procedures for the detection of viroid RNA and certification of disease-free potato tubers. *Phytopathology* **67,** 145–156.

Moutschen, J. (1985). Introduction to Genetic Toxicology. pp. 1–10. J. Wiley, New York.

Murashige, T. (1974). Plant propagation through tissue culture. *Annu. Rev. Plant Physiol.* **25,** 135–166.

Noordham, D. (1973). Identification of Plant Viruses: Methods and Experiments. Center for Agricultural Publishing and Documentation, Wageningen.

Owens, R. A., and Diener, T. O. (1987). Sensitive and rapid diagnosis of potato spindle tuber viroid disease by nucleic acid hybridization. *Science* **213,** 670–672.

Perlman, D. (1979). Use of antibiotics in cell culture media. *In* "Methods in Enzymology, Volume LVIII" (W. B. Jakoby and I. Pastan, eds.), pp. 110–116. Academic Press, New York.

Querfurth, G., and Paul, H. C. (1979). Protein A-coated latex-linked antisera (PALIAS): new reagent for a sensitive test permitting the use of antisera unsuitable for the latex test. *Phytopathologische Zeitschrift* **94,** 282–285.

Redenbaugh, K. (1986). Somatic seeds: encapsulation of asexual plant embryos. *Biotechnology* **4,** 797–801.

Schaad, N. W. (1980). Laboratory Guide for Identification of Plant Pathogenic Bacteria. American Phytopathological Society, St. Paul, Minnesota.

Scharf, S. J., Horn, G. T., and Erlich, H. A. (1986). Direct cloning and sequence analysis of enzymatically amplified genomic sequences. *Science* **223,** 1076–1078.

Sluis, C. J., and Walker, K. A. (1985). Commercialization of plant tissue culture. *IAPTC Newsletter* **47,** 2–12.

Stead, D. E. (1988). Identification of bacteria by computer assisted fatty acid profiling. *Acta Hortic.* **225,** 39–46.

Whitaker, A. M. (1975). Subculture of a cell line: initiation of monolayer cultures in bottles. *In* "Laboratory Manual of Cell Biology" (D. O. Hall and S. E. Hawkins, eds.), pp. 1–4. English University Press, London.

CHAPTER **11**

Controlled Environments in Conventional and Automated Micropropagation

Toyoki Kozai

Laboratory of Horticultural Engineering
Department of Horticulture
Chiba University
Matsudo, Chiba, Japan

I. INTRODUCTION

Development of automated systems for reducing manual operation and for control of the environment in culture is essential to achieving a drastic reduction in production costs for a large-scale micropropagation process.

Recently, much research has been conducted on automation and/or robotization for reducing manual operation in micropropagation, including automation of liquid medium preparation and feeding, plant image recognition and processing, microcutting, and transplanting. On the other hand, not much research has been done on the automation of environmental controls during micropropagation. This is partly because conventional culture vessels are generally so small and airtight that automation in measurement and control of the environment is difficult to conduct.

It is known, however, that many environmental factors affect growth, development, and morphogenesis of plants in culture (Huges, 1981). In addition the culture environment, which is often abnormal when compared with the open field, causes physiological and morphological plant disorders (Debergh and Maene, 1984).

Read (1990) has discussed environmental effects in conventional micropropagation with special attention to the environmental effect during the preculture and culture stages. The reader is also referred to the reviews by Dunstan and Turner (1984) and Kozai (1991a) for the environmental effects on growth and development of micropropagated plants during the acclimatization stage.

This article will review: (1) methods of measurement and control of the environment in culture, (2) effect of *in vitro* environment on photosynthetic growth and development of chlorophyllous tissues, organs, and plants in culture during the multiplication and rooting stages, and (3) control of the *in vitro* environment for multiplication and *in vitro* rooting in automated micropropagation systems. Special attention is paid to environmental factors affecting photosynthesis and the growth of plants during the multiplication and *in vitro* rooting stages.

II. ENVIRONMENTAL FACTORS IN CULTURE

Major environmental factors in culture can be classified as shown in Fig. 1. Not only the average value of each environmental factor but also its variation in time and space affects the growth, development, and morphogenesis of plants. In turn, the variation of the *in vitro* environment is affected by the presence of plants in culture and the medium and by the environment outside the culture vessels.

Thus, the mechanism of environmental changes in a culture vessel is

- *In vitro* environment
  - Aerial physical environment
    - Temperature
    - Light
      - Photosynthetic
      - Photomorphogenic
    - Gaseous composition
      - Carbon dioxide
      - Oxygen
      - Water (humidity)
      - Ethylene
      - Other gases
    - Gas diffusivity (air movement)
    - Pressure
  - Root zone environment
    - Physical environment
      - Temperature
      - Water potential
        - Osmotic, pressure, and matric
      - Gas and liquid diffusivity in the medium
      - Hardness or compactness of the medium
    - Chemical environment
      - Inorganic substance composition
      - Organic substance composition
      - pH
      - Dissolved oxygen
    - Biological environment
      - Symbiotic microorganism
      - Competitive association

Fig. 1. Classification of the environmental factors in culture or the *in vitro* environmental factors which are external to the culture system *per se* but which have an effect on growth, development, and morphogenesis of plants in culture.

similar to that of the environmental changes in a greenhouse. That is, the interrelationship among environments inside and outside the culture vessel and plants in the culture vessel is similar to the interrelationship among environments inside and outside the greenhouse and crops in the greenhouse. In fact, in a sense, a culture vessel is a miniature greenhouse or a small growth chamber. Also, an explant to be cultured *in vitro* may be thought of as a miniature cutting in conventional vegetative propagation (Read, 1990).

Hence, ecological, ecophysical, ecophysiological, environmental physiological, and environmental engineering studies on micropropagation processes should be important. In reality, however, micropropagation research has too long ignored environmental factors and has long relied on pharmaceutical levels of exogenous plant growth regulator application based on huge factorial experiments (Wilkens, 1988).

Environmental control during micropropagation is considered to be important from a practical point of view for the reduction of production costs. Because, as will be shown later, it promotes growth and development (increases in fresh and dry weights, number of nodes, total leaf area, *in vitro* rooting, *in vitro* branching, etc.), it reduces morphological and physiological disorders such as vitrification (also called glassiness) and thin cuticular wax formation on leaves and loss of plants in culture due to biological contamination. It also increases the uniformity of growth and development and reduces excess application of exogenous plant growth regulators. Furthermore, environmental control in culture may help plants grow more rapidly and vigorously *ex vitro* (after transfer to outside conditions).

There are several ways of increasing the number of plants during micropropagation. They include: (1) growth of a single shoot with several nodes from a nodal microcutting without the addition of plant growth regulators in the culture medium, (2) induction of axillary branching with the addition of cytokinins to the culture medium, and (3) induction of direct or indirect regeneration of somatic embryos or adventitious shoots from cells, callus, or tissues, with the addition of plant growth regulators.

No matter which method of increasing the number of plants is used, the main objective of conventional micropropagation is to get a large number of genetically identical plants in a short time period and at a low production cost under predominantly heterotrophic or photomixotrophic conditions, relying on sucrose in the medium as the main carbon source. Then, the plantlets are carefully converted by environmental control to the ones with a higher photosynthetic or photoautotrophic ability (getting carbon dioxide as carbon source from the air) under stressful environmental conditions.

In culture, the carbon source for growth can be either carbohydrates in the medium, CO_2 in the air, or both. Growth dependent purely upon carbohydrates in the culture medium, purely dependent upon photosynthesis, and dependent upon both carbohydrate and photosynthesis are called heterotrophic, photoautotrophic, and photomixotrophic growth, respectively. Photoautotrophic growth can occur only in chlorophyllous cells, tissues, organs, or plants. The effect of environmental factors on growth, development, and morphogenesis depends largely upon the trophic phase.

Often, important practical research subjects in conventional (hetero- or photomixotrophic) micropropagation are finding an optimum nutrient medium for a particular plant and determining when and how the medium should be applied to explants/plantlets *in vitro.*

Explants and regenerated shoots in culture have been considered to have little photosynthetic ability and require sugar (carbohydrates) as a carbon and energy source. Recent research, however, has revealed that chlorophyllous explants/shoots, in general, have relatively high photosynthetic ability and that they may grow faster in some cases under photoautotrophic conditions than under heterotrophic and photomixotrophic conditions, provided that the physical and chemical environments in the culture are properly controlled for photosynthesis (Kozai, 1990).

Furthermore, there is a possibility that the heterotrophic or photomixotrophic phase of shoot regeneration from embryos or adventitious buds is smoothly converted into a photoautotrophic one by proper environmental control. Therefore, research on the photosynthesis of plants in culture is important in order to develop a better micropropagation system.

III. MEASUREMENT AND CONTROL OF THE ENVIRONMENT

For the measurement of the *in vitro* environment, many plant environmental sensors widely used in the fields of agronomy, horticulture, agricultural meteorology, and ecology can be used. Measurement and control of the plant environments are well described by Langhans (1978) and Hanan (1984). However, for the measurement of the *in vitro* environment, sensors and/or the amount of sampling gas should be as small as possible in order not to disturb the environment.

A. Light

Three systems of light measurement have been developed—photometric, radiometric, and quantum (photon) metric. Flux densities (power per unit area incident on a surface) of these are called, respectively, illuminance (unit: lux), irradiance (unit: $W \cdot m^{-2}$) and photon flux density, or simply photon flux (mol $m^{-2}s^{-1}$), 1 mole being equal to 6.02×10^{23} photons. The values of these three flux densities can only be converted by careful consideration of the complex interrelationships that depend on the sensitivity of various light receptors (McFarlane, 1978).

Photometry specifically deals with the measurement of light seen by the "standard (normal) human eye" so that illuminance is not an appropriate indicator to describe the light environment for plants. Irradiance and photon flux density are, respectively, appropriate indicators to describe the light environment in terms of energy budget and photosynthesis/photomorphogenesis.

The photosynthetic reaction is driven by light in the spectral range between 400 and 700 nm. Irradiance and photon flux density in this wave band are called, respectively, photosynthetically active radiation (PAR) and photosynthetic photon flux density (PPFD), or simply photosynthetic photon flux (PPF). Compact sensors (e.g., 15 mm in diameter and 20 mm in height) detecting PAR and PPFD can be obtained at a reasonable cost.

Photomorphogenesis results from a low level of photon flux density and is not significantly influenced by its intensity. Most photomorphogenic reactions are controlled by wavelengths either in the blue region (400–460 nm) or in the red (620–680 nm) and far-red (700–800 nm) regions. Therefore, sensors detecting the regions 400–460, 620–680 and 700–800 nm for photomorphological research should be sensitive at low levels of PPFD for each wave band.

In short, the light flux density should be measured with a PPFD meter for photosynthetic research. For photomorphological research, however, a photon flux density meter which separately measures in the regions of 400–460, 620–680, and 700–800 nm is required, in addition to the PPFD meter. In research on energy budgets, on the other hand, radiometric sensors such as a PAR sensor and a net radiometer are necessary.

Spectral distributions of light from different light sources significantly differ from each other (Bickford and Dunn, 1972). Fluorescent lamps have been the primary light source used in micropropagation. The spectrum generally matches the requirements of cultures. When the lamps are placed close together, they generally give a relatively uniform horizontal distribution of PPFD on the culture shelf. However, there is a large difference in PPFD between the inside and outside of the vessels, and the distribution of PPFD in culture vessels on the shelf is largely dependent upon the closure type, vessel type, and their arrangement on the shelf (Fujiwara *et al.*, 1989).

The ballast or stabilizer, which produces heat and consumes more than 10% of the total electricity for lighting, is necessary for lighting fluorescent lamps. Recently, a transistor-type inverter has been used in place of a conventional ballast to reduce electricity consumption and hence heat production by 15–20%. The inverter converts direct current to alternating current with 40–50 kHz, thus producing minimum flickering. With this inverter, the output of PAR can be continuously adjusted to the desired value. The inverter can be placed outside the culture room, as with the conventional ballast, to further reduce the electricity consumption for cooling.

The light source is almost always installed above the vessels and the tissues/plants in the vessels generally receive downward light. In this case, as the plants grow, more light energy is intercepted by the upper part of the plants and less light energy by the lower parts. In future micropropagation systems, it might be better to direct light from the sides by means of optical fibers or other thin light sources (Kozai, 1990). With this lateral lighting method, plants may receive more light evenly with less electricity consumption for lighting. The lateral lighting may promote better growth and may give better plant shape. For control of plant photomorphogenesis with a low level of blue, red, and far-red lights, different types of light emitting diodes (LED), which emit either blue, red, or far-red light separately, can be used in the future.

B. Gas Exchange Characteristics of Vessels

It has been shown that the type of vessel closure affects the gaseous composition as well as the light environment and hence vitrification and growth of plants in culture, the looser types being better than the tighter ones for reducing vitrification of plantlets *in vitro* (Hakkaart and Versluijs, 1983; Dillen and Buysens) and for promoting the growth of strawberry plantlets *in vitro* (Kozai and Sekimoto, 1988).

The air exchange characteristics of the vessel are best expressed by the number of air changes (infiltration) per hour of the vessel, which is defined as the hourly air change rate of the vessel divided by the air volume of the vessel (Kozai *et al.*, 1986). The number of natural air changes for a flat bottom glass test tube (air volume: 45 ml) closed with an aluminum foil cap, plastic formed cap, and silicon foam rubber plug are 0.18, 1.5, and 0.6 per hour, respectively (Kozai *et al.*, 1986). The number can be increased to 3–6 times per hour by using a gas-permeable microporous polypropylene film as part of the vessel closure (Kozai and Sekimoto, 1988).

The number of natural air changes per hour of an empty vessel, E, can be estimated using carbon dioxide gas as a tracer gas by the following equation:

$$E = -\ln[(K - K_{ou})/(K_0 - K_{ou})]/T$$

where K and K_0 (ppm or $\mu mol \cdot mol^{-1}$) are the gas concentration inside the vessel at time t and time 0, respectively; K_{ou} is the constant gas concentration outside the vessel; T is the time interval from time 0 to t (hours). Generally, K_{ou} is kept at around 350 ppm (normal atmospheric CO_2 concentration), K and T are in the range of 4,000 to 8,000 ppm and 0.5–2.0 hour, respectively.

E is a sort of physical property of the vessel and is almost constant with time. The gas concentration in the vessel containing shoots or plantlets and medium varies with the gas concentration outside the vessel, the E value and gas production/absorption characteristics of the plants/medium in the vessel. The mathematical relationship among the factors previously mentioned is given by Fujiwara *et al.* (1987). The E can be defined similarly for forced air changes in the vessel.

C. Relative Humidity and Water Potential

The interchange of water in the gaseous and/or liquid phase among plants, aerial and root zone environments in the vessel, and the air outside the vessel plays an important role in plant growth and development. Directions and rates of water flow are determined by the spatial distribution of water potentials inside and outside of the vessel; water flows from points with higher water potential values to points with lower water potential values. Water potential of the air, P_w (unit: Pa; 10 bar = 10^6 Pa = 1 MPa), can be calculated by the following equation:

$$P_w = -(R \cdot T/v) \cdot \ln(H/100)$$

where R is the perfect gas constant ($8.31\ J \cdot mol^{-1} \cdot K^{-1}$), T is absolute temperature (K), v is the partial molal volume of water ($1.8 \cdot 10^{-5} m^3 \cdot mol^{-1}$), and H is the relative humidity (percent). The water potential is zero for air saturated with water vapor (relative humidity being 100%).

Relative humidity is normally high in the vessel. The high relative humidity in the vessel can be rather accurately measured with a high polymer sensor (impedance type) if calibration is conducted using saturated salts at a regular time interval. The compact cylinder type sensor (8 mm in diameter and 35 mm in length) is commercially available at a reasonable cost. Most humidity sensors are attached with a thermistor for temperature measurement.

During the photoperiod when a cooling unit is often operated, the relative humidity in culture room is generally low (30–40%) due to dehumidification by the cooling unit. When vessels with a relatively high number of air changes are used, excess evapotranspiration may occur. An ultrasonic humidifier for home use can be used to keep relative humidity at a certain high level. Relative humidity can be rather accurately controlled by using a humidifier with a humidistat.

Kozai *et al.* (1990) showed the dependence of relative humidity in the vessel on the number of air changes per hour, relative humidity in the culture room, and total leaf area of the plantlets in the vessel, based upon relative humidity measurements using the high-polymer sensors.

The water potential of liquid under normal atmospheric pressure conditions is equal to a negative osmotic pressure, otherwise called the osmotic potential. (The water potential of pure water is zero.) The water potential of a gelled medium under normal atmospheric pressure condition is a sum of the osmotic potential and matric potential, the matric potential being normally small compared with the osmotic potential.

The water (osmotic) potential of liquid media is practically the sum of the water potentials due to the basic component (inorganic matter) and sugar (carbon source and some agents for osmotic potential regulation such as mannitol and sorbitol).

The water potentials due to the basic components of some widely used culture media are −212 kPa for Murashige and Skoog (1962), −89 kPa for Heller (1953), −37 kPa for White (1962), −143 kPa for Gamborg, Miller, and Ojima (1968), −106 kPa for Nitsch (1969) and Schenk and Hildebrandt (1972) (Kozai *et al.*, 1986).

The water potential due to sugar can be estimated by multiplying −0.78 by its percent (weight/volume) for the disaccharide such as sucrose and multiplying −1.46 by its percent (weight/volume) for monosaccharide such as glucose, fructose, mannitol, and sorbitol (Kozai *et al.*, 1986).

The water potential of liquid media and sap of plant tissue can be measured with an osmometer. The water potential of gelled media and plant tissue can be measured with a psychrometer. The water potential of plant tissue is the sum of the osmotic potential, the matric potential, and the pressure potential (turgor pressure).

D. CO_2, O_2, and Ethylene

Concentrations of CO_2, O_2, and ethylene (C_2H_4) can be measured simultaneously using a gas chromatograph with an FID (flame ionized detector). The amount of sample gas required for the measurement is in the range of 0.25 ml to 1 ml.

CO_2 concentration is also measured with an infrared-type CO_2 analyzer. This is convenient for a continuous measurement, but the analyzer requires a relatively large amount of sample gas for the continuous measurement, and the sampling may disturb the gaseous environment. O_2 concentration is also measured with an oxygen analyzer having a zirconium oxide detector. Approximately 5 ml of sampling gas is required for one O_2 measurement with the analyzer. Ethylene can be easily measured using only a gas chromatograph with an FID.

There are several ways of modifying the gaseous components in the vessel. They include: (1) addition of agents absorbing or producing a particular gas component in the vessel, (2) control of the gaseous environment in the culture room and use of a gas-permeable film as closures, and (3) use of a forced ventilation or air circulation system.

1. Addition of agents. To reduce the concentrations of CO_2 and ethylene in the vessel, soda lime and $KMnO_4$ are used as absorbents, respectively. O_2 concentration can be reduced by an O_2 absorbent which is widely used for reducing bacterial and fungal contamination during fresh food storage.
2. Control of the gaseous environment in the culture room. CO_2 concentration in the culture room can be controlled relatively easily using an infrared-type CO_2 controller with a pure CO_2 container. Relative humidity can be controlled using a humidistat with a dehumidifier and humidifier. It is difficult to control O_2 concentration in the culture room.
3. Forced ventilation system. There are several types of forced ventilation systems for controlling the gaseous environment in the vessel (e.g., Fujiwara *et al.*, 1988; Walker *et al.*, 1988, Cuello *et al.*, 1989). In a large-scale micropropagation system, the forced ventilation system will become more and more important.

E. CO_2 Concentration in the Vessel during the Photoperiod

The decrease in CO_2 concentration in the culture vessel during the photoperiod (C_{in}) was first reported by Ando (1978) for orchids being cultured *in vitro* at the rooting stage.

The C_{in} in relatively airtight vessels containing ornamental plantlets decreased to 70–80 ppm within 2–3 hours after the start of the photoperiod (Fujiwara *et al.*, 1987; Infante *et al.*, 1989). The C_{in} was as low as the CO_2 compensation point of C_3 plants and was about 250 ppm lower than the normal atmospheric CO_2 concentration (ca. 340 ppm).

Experimental results suggest the following for chlorophyllous explants/plantlets being cultured *in vitro* by conventional micropropagation: (1) The explants/plantlets have a photosynthetic ability because C_{in} decreases sharply with time at the onset of the photoperiod. (2) Their photosynthesis is restricted by low CO_2 concentrations during most of the photoperiod because there is insufficient CO_2 from the outside air through the vessel caps. (3) They are forced to develop hetero- or photomixotrophy, and a higher PPF will not increase net photosynthetic rates under such low CO_2 conditions. (4) They may develop photoautotrophy and grow faster under photoautotrophic, high CO_2, and high PPF conditions than under hetero- or photomixotrophic conditions. (5) The initial growth rate is greater for an explant with a larger area of highly chlorophyllous tissue (green leaves) under high CO_2 and light conditions.

F. Net Photosynthetic Rate

Estimation of the net photosynthetic rate of tissue-cultured explants/plantlets is mostly conducted using a small assimilation chamber with a forced air mixing system. CO_2 measurements are made with an infrared-type CO_2 analyzer which requires a significant volume of sample gas (e.g., Grout and Ashton, 1978; Donnelly and Vidaver, 1984; Reuther, 1988; Capellades, 1989). In this case, the net photosynthetic rate per plantlet is estimated by multiplying the forced air flow rate through the chamber by the difference in CO_2 concentration between the inlet and outlet of the chamber. The difference is relatively small even at a low air flow rate because the chlorophyllous parts of the explants/plantlets and hence CO_2 uptake in the chamber is small.

Furthermore, with the assimilation chamber method, physical parameters such as the gas diffusion coefficients adjacent to the plantlets and the physiological conditions such as water relations of the explants/plantlets in the chamber may differ from those *in vitro* and *in situ* (in the vessel). The net photosynthetic rates of *in vitro* explants/plantlets *in situ* may then differ from those in the chamber for identical PPF, CO_2, and temperature levels. We must be careful, therefore, to interpret the results obtained with assimilation chambers.

Fujiwara *et al.* (1987) showed a numerical method for estimating the parameter values of photosynthetic response curves of *in vitro* plantlets *in situ* using the data on the diurnal courses of CO_2 concentration inside and outside the vessel. Kozai and Iwanami (1988) estimated the steady-state net photosynthetic rate of *in vitro* plantlets *in situ* by multiplying the difference in CO_2 concentration between the inside and outside of the vessel by the natural air ventilation rate, hourly natural ventilation rate being a product of the number of air changes per hour of the vessel, the air volume of the vessel, and a unit conversion factor. In these two methods, CO_2 measurements are made using a gas chromatograph with a negligible disturbance, allowing the net photosynthetic rate of *in vitro* plants to be estimated *in situ*. A similar method can be employed for estimating the dark respiration rate, transpiration rate, etc.

IV. ENVIRONMENTAL EFFECT ON PHOTOSYNTHETIC GROWTH AND DEVELOPMENT

In photoautotrophic micropropagation, the photosynthesis, growth, and development of *in vitro* explants/plantlets are largely influenced by the physical environmental factors which include light, CO_2, humidity, airflow speed, temperature, and O_2. Research on the environmental effect on photoautotrophic micropropagation has been extensively reviewed by Kozai (1990, 1991). In the following sections, the environmental effect on the net photosynthetic rate, growth, and development of explants/plantlets *in vitro* will be briefly reviewed.

A. Photosynthetic Response Curves of Explants/Plantlets *in Vitro*

Kozai *et al.* (1990) showed the photosynthetic response curves for *in vitro Cymbidium* plantlets *in situ* as affected by CO_2, PPF, and temperature. They showed that the curves did not differ largely from those of plants grown under shade in the greenhouse.

The net photosynthetic rate of *in vitro Primula malacoides* (C_3 plant) plantlets in 1% O_2 and 10% O_2 were, respectively, about 3 times and 1.5 times larger than that in 21% O_2, at a CO_2 concentration of approximately 200 ppm, due to a reduced photorespiration at lower O_2 concentrations (Shimada *et al.*, 1988).

The net photosynthetic rate of *in vitro* rose plantlets was greater when cultured on the medium with a lower sucrose concentration (Capellades, 1989). A similar result was obtained for potato plantlets *in vitro* (Kozai *et al.*, unpublished observation). The starch content in leaves of plantlets was higher when cultured on the medium with a higher sucrose concentration, while a higher starch content in the leaves brought a lower net photosynthetic rate (Capellades, 1989). It was also shown that the rate was affected by relative humidity.

The net photosynthetic rate of strawberry plantlets was greater when cultured in a vessel with a forced ventilation system than when cultured in a normal vessel with natural ventilation (Kozai *et al.*, 1989). This is probably because CO_2 diffusion into the stomatal cavities of the plantlets *in vitro* is restricted by the stagnant air in the normal vessel, the air movement in the vessel being caused by natural convection.

The net photosynthetic rates of leaves of plantlets and seedlings *in vitro* measured under saturated PPF, 340 ppm CO_2, and a leaf temperature of 20°C were similar. Nevertheless the shape of the photosynthetic response curves relating the rate and CO_2 concentration indicated some differences in their photosynthetic parameters (Pospisilova *et al.*, 1987).

Research on photosynthetic responses of *in vitro* explants/plantlets *in situ* as affected by the *in vitro* physical environment has only been recently conducted, and many problems still need to be solved.

B. CO_2 Enrichment under High PPF

Based upon the data given above, one can expect an increase in net photosynthetic rate and hence growth of plantlets *in vitro* by increasing the CO_2 concentration in the vessel during the photoperiod. Practical ways of increasing CO_2 concentration include:

1. Use of gas-permeable film as closures.
2. Use of gas-permeable film as closures in CO_2-enriched culture room.
3. Use of spaghettilike tubes and an air pump with an air filter for forced ventilation of conventional culture vessels.
4. Use of a larger vessel with a CO_2 supply system or a forced ventilation system.

With these systems, not only CO_2 concentration but also relative humidity, ethylene, gas diffusivity, etc. in the vessel are modified. Therefore, the changes in growth of explants/plantlets *in vitro* resulting from the use of these systems cannot be attributed only to CO_2 enrichment. However, the changes are considered to be brought about primarily by the CO_2 enrichment in most cases. More detailed studies on the effect of a gaseous environment on the photosynthetic growth and development is an area for further study.

1. Gas-Permeable Film in Closures

There are several reports indicating the positive effects of the use of gas-permeable film as closures under high PPF on increases in net photosynthetic rate and growth of plantlets *in vitro* (Kozai, 1990). When leafy single-node cuttings are used as explants, the growth of plantlets of some species is greater when cultured on medium without sugar in the vessel closed with the film than when cultured on medium with sugar in the relatively air-tight vessel. With this approach, the effects of the passive CO_2 enrichment under high PPF on growth *in vitro* can be verified with a minimum change in the existing micropropagation facility. The percentage of vitrification is also decreased with the use of the film, possibly because of lower humidity and increased gas exchange in the vessel.

2. CO_2 Enrichment in the Culture Room

CO_2 enrichment under high PPF (100–200 μmol $m^{-2}s^{-1}$) was effective for promoting the chlorophyllous explant/plantlet growth of tobacco (Mosseau, 1986), *Cymbidium* (Kozai *et al.*, 1987), carnation (Kozai and Iwanami, 1988), and potato (Kozai *et al.*, 1988) when cultured on the medium both with and without sugar.

3. A Larger Vessel with a CO_2 Supply System

Forced ventilation with atmospheric air or nitrogen–oxygen–carbon dioxide mixtures resulted in reduced propagule weights and shoot numbers for Stage II micropropagation of *Rhododendron* in the vessels with 400 ml air volume when cultured under a PPF of 39 μmol $m^{-2}s^{-1}$ (Walker *et al.*, 1988).

The dry weight and net photosynthetic rate of strawberry plantlets cultured with sugar-free liquid medium were promoted when cultured in a large vessel with a forced ventilation system under a PPF of 96 μmol $m^{-2}s^{-1}$, compared with those cultured by a conventional method (Fujiwara *et al.*, 1988).

Much more research should be done in future on the effects of forced ventilation on the growth of explants/plantlets.

V. ADVANTAGES OF PHOTOAUTOTROPHIC MICROPROPAGATION

Hetero- or photomixotrophic micropropagation has many advantages over conventional vegetative propagation. However, its widespread com-

TABLE I
Some Disadvantages of Hetero- and Photomixotrophic Micropropagation

1. Sugar in the medium may cause biological contamination.
2. Airtight, small vessels must then be used to reduce a loss due to contamination.
3. Automated, robotized, and computerized micropropagation system is thus difficult to develop.
4. The inside air is almost saturated with water vapor.
5. CO_2 and ethylene concentrations may become abnormal.
6. High PPF is not effective for growth promotion.
7. Growth regulators are often necessary for regeneration.
8. The abnormal environment may induce physiological/morphological disorders, retardation of plantlet growth, and mutation.
9. The disorders may result in a high percentage of death during the acclimatization stage.

mercial use is still restricted as a result of its relatively high production costs. Some disadvantages of hetero- and mixotrophic micropropagation are shown in Table I.

Photoautotrophic micropropagation seems to have many advantages for reducing the production costs and for increasing productivity and plantlet quality (see Table II). Some of the advantages shown in Table II have been proven in experiments, but many have not been and require additional research, especially the environmental effects on photosynthetic growth of explants/plantlets *in vitro*.

However, the potential of photoautotrophic micropropagation using chlorophyllous shoots/nodal cuttings, whether automated or not, seems promising.

TABLE II
Some Advantages of Photoautotrophic Micropropagation

1. Growth and development of plantlets are promoted.
2. Physiological/morphological disorders and mutation are reduced, and plantlet quality is improved.
3. Procedures for rooting and acclimatization are simplified.
4. Application of growth regulators and other organic matter can be minimized.
5. A loss of plantlets due to contamination is reduced.
6. A larger vessel can be used with minimum contamination.
7. Then, the environmental control of the vessel is easier.
8. The control of growth and development by means of environmental control is easier.
9. Automation, robotization, and computerization are easier.

VI. CONCLUDING REMARKS

In this chapter, methods of measuring and controlling the *in vitro* environment are given. It has also been shown that the chlorophyllous explants/plantlets have photosynthetic ability and they may grow well on a medium without sugar, provided that the environment is properly controlled for promoting photosynthetic growth.

In photoauto- and photomixotrophic micropropagation, the environmental control for optimizing the photosynthetic growth and development is essential, just as in conventional vegetative micropropagation and seedling production.

Based on these considerations, a unified principle and a basic research area of the environmental effects on transplant production can be established in general. The transplants include aseptically micropropagated and septically propagated plantlets and seedlings. The production of transplants using cuttings and seeds is normally conducted photoautotrophically; therefore, there should be some common principles and many common techniques in the production of different kinds of transplants.

REFERENCES

Aitken-Christie, J. (1990). Automation. *In* "Micropropagation of horticultural crops." (P. Debergh and R. Zimmerman, eds.) Kluwer Academic Publishers, Dordrecht (in press).

Ando, T. (1978). Gaseous environment in the airtight culture vessel containing orchids. *Abstr. Annual Autumn Meet. Jap. Soc. Hort. Sci.* 368–369.

Bickford, E. D., and Dunn, S. (1978). "Lighting for Plant Growth." The Kent University Press, Ohio. 221 pp.

Capellades, M. Q. (1989). Histological and ecophysical study of the changes occurring during the acclimatization of *in vitro* cultures. Ph.D. diss., pp. 98. Gent University, Belgium.

Cuello, J. L., Walker, P. N. and Heuser, C. W. (1989). Effects of ventilated CO_2, light and sucrose levels on stage II micropropagation of *Buddleia alternifolla* (Butterfly Bush). ASAE Paper No. 89-6091. (ASAE/CSAE presentation paper).

Debergh, P., and Maene, L. (1984). Pathological and physiological problems related to the *in vitro* culture of plants. *Parasitica* **40,** 69–75.

Dillen, W., and S. Buysens (1989). A simple technique to overcome vitrification in *Gypsophila paniculata* L. *Plant Cell, Tissue Organ Cult.* **19,** 181–188.

Donnelly, D. J., and Vidaver, W. E. (1984). Pigment content and gas exchange of red raspberry *in vitro* and *ex vitro*. *J. Amer. Soc. Hort. Sci.* **109,** 177–181.

Dunstan, D. I., and Turner K. E. (1984). The acclimatization of micropropagated plants. *In* "Cell Culture and Somatic Cell Genetics of Plants." Vol. 1. (I. K. Vasil, ed.), 123–129. Academic Press, Orlando.

Fujiwara, K., Kozai, T., and Watanabe, I. (1987). Measurements of carbon dioxide gas concentration in closed vessels containing tissue cultured plantlets and estimates of net photosynthetic rates of the plantlets. *J. Agric. Meteorol.* **43,** 21–30. (in Japanese with English summary).

Fujiwara, F., Kozai, T., Nakajyo, Y., and Watanabe, I. (1989). Effects of closures and vessels on light intensities in plant tissue culture vessels. *J. Agric. Meteorol.* **45 (3),** 143–149. (in Japanese with English summary).

Grout, B. W. W., and Ashton, M. J. (1978). Transplanting of cauliflower plants regenerated from meristem culture. II. Carbon dioxide fixation and the development of photosynthetic ability. *Hort. Res.* **17,** 65–117.

Infante, R., Magnanini, E., and Righetti, B. (1989). The role of light and CO_2 in optimizing the conditions for shoot proliferation of *Actinidia deliciosa in vitro. Physiol. Plantarum.* **77,** 191–195.

Hakkaart, F. A., and Versluijs, J. A. (1983). Some factors affecting glassiness in carnation meristem tip cultures. *Neth. J. Pl. Path.* **89,** 47–53.

Hanan, J. J. (1984). "Plant Environmental Measurement." Bookmakers Guild, Colorado. 326 pp.

Hughes, K. W. (1981). *In vitro* ecology: Exogenous factors affecting growth and morphogenesis in plant culture systems. *Environ. Exper. Botany.* **21,** 281–288.

Kozai, T., K. Fujiwara, and I. Watanabe (1986). Relation between the culture medium composition and water potential of liquid culture media. *J. Agric. Meteorol.* **42,** 1–6. (in Japanese with English summary).

Kozai, T., Fujiwara, K., and Watanabe, I. (1986). Effects of stoppers and vessels on gas change rates between inside and outside of vessels closed with stoppers. *J. Agric. Meteorol.* **42,** 119–127. (in Japanese with English summary).

Kozai, T., Oki, H., and Fujiwara, K. (1987). Effects of CO_2 enrichment and sucrose concentration under high photosynthetic photon fluxes on growth of tissue-cultured *Cymbidium* plantlets during the preparation stage. *Proc. Symposium Floriezel,* **87,** Arlon - Belgium. 135–141.

Kozai, T., and K., Sekimoto (1988). Effects of the number of air changes per hour of the closed vessel and the photosynthetic photon flux on the carbon dioxide concentration inside the vessel and the growth of strawberry plantlets *in vitro. Environ. Control Biol.* **26,** 21–29.

Kozai, T., and Iwanami, Y. (1988). Effects of CO_2 enrichment and sucrose concentration under high photon fluxes on plantlet growth of carnation (*Dianthus caryophyllus* L.) in tissue culture during the preparation stage. *J. Japan. Soc. Hort. Sci.* **57(2),** 279–288.

Kozai, T., Koyama, Y., and Watanabe, I. (1988). Multiplication of potato plantlets *in vitro* with sugar free medium under high photosynthetic photon flux. *Acta Hortic.* **230,** 121–127.

Kozai, T., Kubota, C., and Nakayama, M. (1989). Net photosynthetic rates of plantlets *in vitro* under natural and forced ventilation conditions. *J. Jap. Soc. Hort. Sci.* Vol. **58.** Special issue No. 1. 250–251.

Kozai, T. (1990). Micropropagation under photoautotrophic conditions. *In* "Micropropagation." (P. Debergh, and R. H. Zimmerman, eds.) 449–471. Kluwer Scientific, Dordrecht.

Kozai, T. (1991a). Acclimatization of micropropagated plants. *In* "Biotechnology in Agriculture and Forestry." Vol. **17,** (Y. Bajaj, ed.) Springer-Verlag, New York. (in press).

Kozai, T. (1991b). Autotrophic micropropagation. *In* "Biotechnology in Agriculture and Forestry." Vol. **17,** (Y. Bajaj, ed.) Springer-Verlag, New York. (in press).

Kozai, T., Tanaka, K., Watanaba, I., and Hayashi, M. (1989). Measurement of relative humidity and transpiration characteristics of plantlets *in vitro*. *Abst. of Annual Meet. of Agric. Meteorol.*, 24–25. (in Japanese).

Kozai, T., Oki, H., and Fujiwara, K. (1990). Photosynthetic characteristics of *Cymbidium* plantlet *in vitro*. *Plant Cell, Tissue Organ Cult.* **22,** 205–211.

Langhans, R. W. (Ed.) (1978). "A Growth Chamber Manual." 222 pp. Comstock Pub., Associates, Ithaca.

McFarlane, J. C. (1978). Light *In* "A Growth Chamber Manual (Environmental Control for Plants)" (R. W. Langhans, ed.) pp. 15–44. Comstock Pub. Associates, Ithaca.

Moe, R. (1988). Effect of stock plant environment on lateral branching and rooting. *Acta Hortic.* **226,** 431–444.

Mousseau, M. (1986). CO_2 enrichment *in vitro*. Effect on autotrophic and heterotrophic cultures of *Nicotiana tabacum* (Var. *Samsun*). *Photosynth. Res.* **8,** 187–191.

Pospisilova, J., Catsky, J., Solarova, J., and Ticha, I. (1987). Photosynthesis of plant regenerants. Specificity of in vitro conditions and plantlets response. *Biologia plantarum (Praha).* **29,** 415–421.

Pospisilova, J., Solarova, J., Catsky, J., Ondrej, M., and Opatrny, Z. (1988). The photosynthetic characteristics during the micropropagation of tobacco and potato plants. *Photosynthetica* **22,** 205–213.

Read, P. E. (1989). Environmental effects in micropropagation. *In* "Handbook of plant cell culture" Vol. 5. (P. V. Ammirato, D. A. Evans, W. R. Sharp, and Y. P. S. Bajaj, eds.). pp. 95–125. McGraw-Hill Pub., New York.

Reuther, G. (1988). Comparative anatomical and physiological studies with ornamental plants under in vitro and greenhouse conditions. *Acta Hortic.* **226,** 91–98.

Walker, P. N., Heuser, C. W., and Heinemann, P. H. (1988). Micropropagation: Studies of gaseous environments. *Acta Hortic.* **230,** 145–151.

Wilkins, H. F. (1988). Techniques to maximize cutting production. *Acta Hortic.* **226,** 137–143.

CHAPTER **12**

The Use of Robotics in Automated Plant Propagation

Nobuyuki Fujita
Atsushi Kinase

Mechanism and Control Technology Department
Manufacturing Engineering Laboratory
Toshiba Corporation
Yokohama, Japan

I. INTRODUCTION

The use of tissue culture techniques for the mass production of plantlets has increased substantially during the past two decades. There are many reasons for the growth of this industry. Modern methods have been used to develop many new varieties of ornamental, foliage, and vegetable plants which bear no seeds because of the lack of fertility. In those instances where seeds are formed, they often do not produce true-to-type plants because of genetic segregation. Tissue culture methods are very useful in these instances for the clonal and rapid propagation of the improved plants. Millions of plants are produced by well-established micropropagation techniques in tissue culture laboratories around the world. Industrial-scale mass production of high-quality plantlets and their supply to cultivators increases the efficiency and value of their businesses. Large investments of time and financial resources are normally required for these operations.

For these and the many other reasons discussed in various chapters of this volume, the production and marketing of plants produced in tissue culture has increased. However, further expansion of the market is limited by the high cost of the plants produced, the primary reason for which is that most of the operations in this process are still performed manually. Among all the processes of tissue culture propagation, cutting, and transplant operations, which are necessary for the rapid multiplication of plantlets, cutting-off and transplanting are two of the most labor intensive.

Automation by robots has been proposed as an effective method for low-cost multiplication and mass production of plants by tissue culture techniques. In this chapter we describe the development and operation of robots for automated plant propagation.

II. PLANTLET PRODUCTION THROUGH TISSUE CULTURE

The outline of a plantlet production system through tissue culture is shown in Fig. 1. First, an excellent parent tissue explant, such as an apical meristem, is trimmed and placed on culture medium to induce either the development of axillary buds or in some cases callus. The use of axillary buds is more common and preferable. However, in those instances in which a callus is formed, adventitious shoots can be induced by providing appropriate plant growth regulators. The shoots formed *in vitro* are transferred to other media to obtain rooted plantlets. For rapid multiplication, however, the shoots are divided and retransplanted on

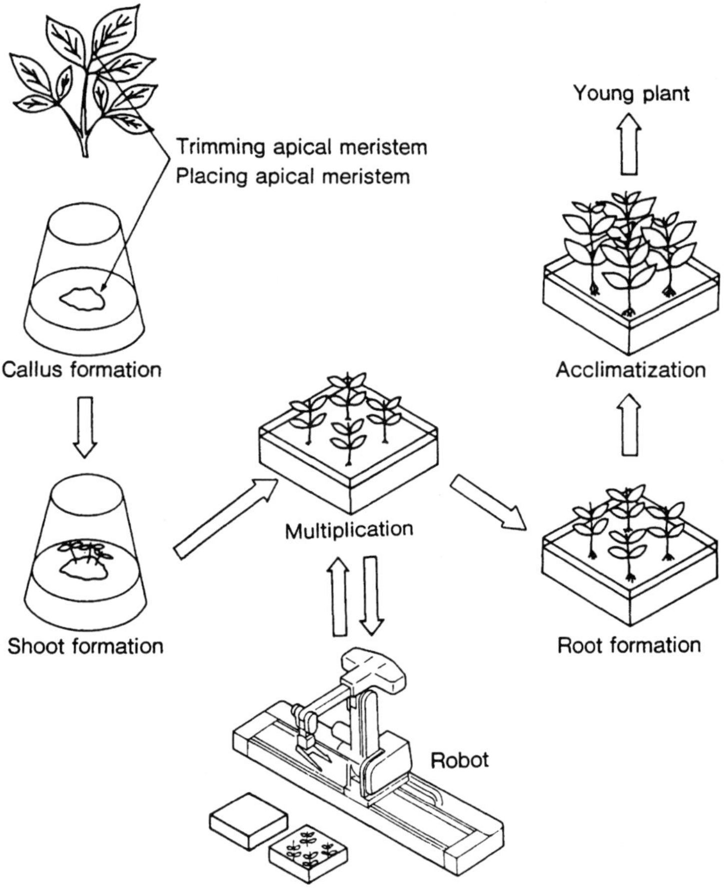

Fig. 1. Process for plantlet production.

the initial medium which supports the development of axillary buds. This procedure can be performed repeatedly to obtain millions of plants. In this chapter, we describe the automation of the division and transplanting of shoots.

During the division of the plantlets it is essential that each cut-off portion has more than one node. From each node a new axillary bud is formed, which then can be developed into a shoot.

Apical buds are commonly used as the primary explants for micropropagation because they are often free from viruses. The subsequent operations can then be performed in virus-free conditions. The automated operations of cutting off and transplanting by robots have the advantage of preventing virus contamination as well as saving labor.

III. NECESSITY OF ROBOTS AND REQUIRED FUNCTIONS

A. Purposes of Automation

Tissue cultured plantlets are expensive because of the high labor costs involved in their production. Automation of the cutting and transplanting operations can provide significant reductions in the amount of labor used for these time-consuming but routine functions. Automation can also enhance the productivity of the system and the quality of the product. Fatigue, mistakes, and boredom, which are serious problems for manual labor, can be entirely avoided by the use of robots which can work 24 hours a day. Furthermore, the ever present risks of contamination can be reduced by introducing a disinfection process into the robotic system and by providing for the automatic detection and discarding of defective and infected plants.

B. Characteristics Required for the Robots

In order to handle tender living plants, the operation and capabilities of the robots must be significantly different from those of conventional industrial robots. Because plantlets are not solid and may vary greatly in shape and size, most operations by human hands are performed based on the judgment of the individual operator as to where to grip and where to cut. In order to replace the human operators, highly intelligent and capable robots must be developed.

The robots must have the following functions and capabilities:

1. The objects to be handled by the robots have different shapes and dimensions, and their positions are not fixed. Therefore, suitable methods of detecting the three-dimensional shapes, dimensions, and positions of the objects are necessary. Otherwise, at least the key parts of the objects must be guided to fixed positions by some means.
2. The functions of judging how many portions to cut, where to grip, where to cut, and so on must be provided.
3. As the objects are tender and living, the robots must be able to grip them softly and safely without inflicting mechanical damage.

4. As described above, the plantlets must be free of virus contamination and other pathogens. Therefore, not only must the robots be easily sterilized, but also effective measures must be taken against the spread of contaminants to other plantlets in those instances in which some plantlets may be contaminated.
5. The robots should be able to operate faster than human operators.

IV. DEVELOPMENT OF THE PROTOTYPE ROBOT SYSTEM

A. Overview

The prototype robot was developed in May 1988 and was exhibited at the "High-Tech Horti-mation '88" fair in Tokyo. This robot was developed to demonstrate the possibility of automated cutting and transplanting operations and to evaluate the necessary functions for them. The development was based on the principle of using, as far as possible, previously developed technology and parts.

Color Plate 10 (Watake and Kinase, 1988) shows this robot system, which consists of the robot itself, a force-controlled gripper (called a soft gripper), a plantlet detection sensor with a cutter (lower right in Color Plate 10), and some trays with plantlet stems in them. This robot can detect the position of stems in a tray (lower center), cut the stems into proper sizes, and transplant them into other trays (left).

Figure 2 shows the total system structure. The microcomputer system, which has Motorola's MC68020 microprocessor, controls the total system as a mainframe computer. The robot itself is controlled by the robot controller, which is supervised by the microcomputer system. On the other hand, the soft gripper and the position detective sensor are directly controlled by the microprocessor.

B. The Robot

This robot had been developed earlier for semiconductor wafer handling and has the following advantages for plant propagation:

1. The robot has 6 degrees of freedom. It can, therefore, basically position any point with any attitude within the movement field.
2. The robot has been designed as a dust-free clean system and is rust-free in moist conditions.

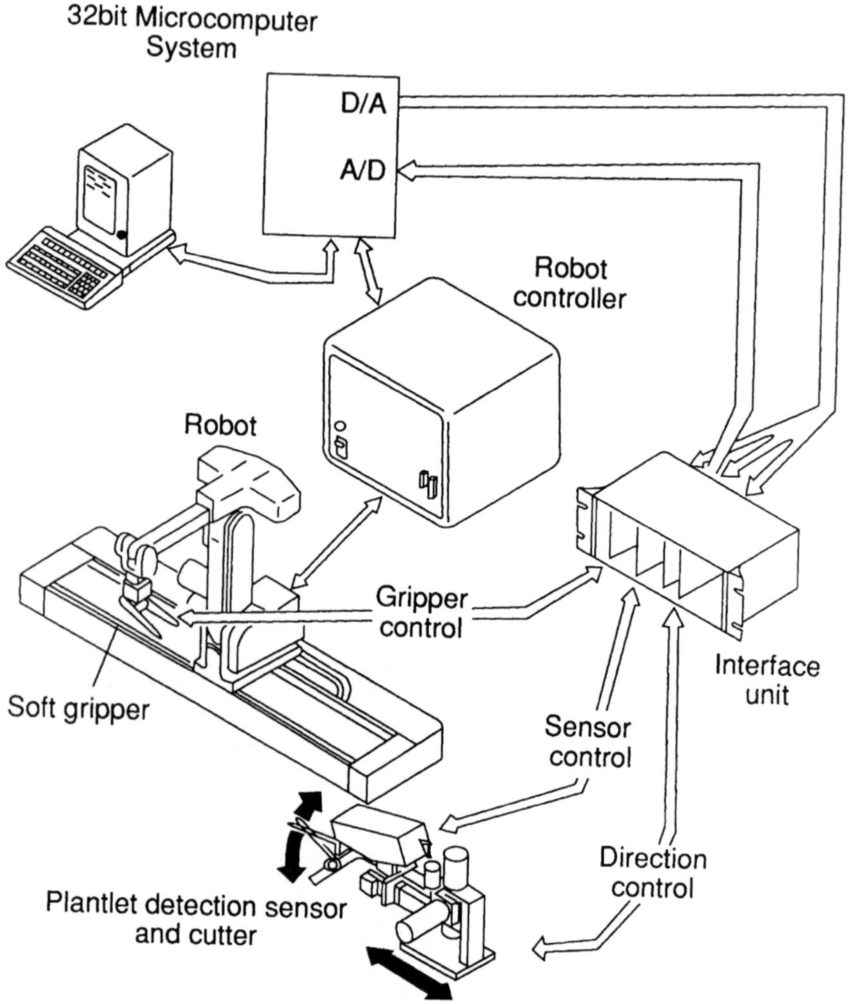

Fig. 2. System structure of the prototype robot system (Watake and Kinase, 1988).

3. The structure of straight movement of the first axis is better for clean room operation because of its wider movement field compared with the robot size.

C. Plantlet Detection Sensor and Cutter

Figure 3 shows the plantlet detection sensor which has three DC servomotors, a laser emitter, a scanning mirror, and a position sensitive detector (PSD).

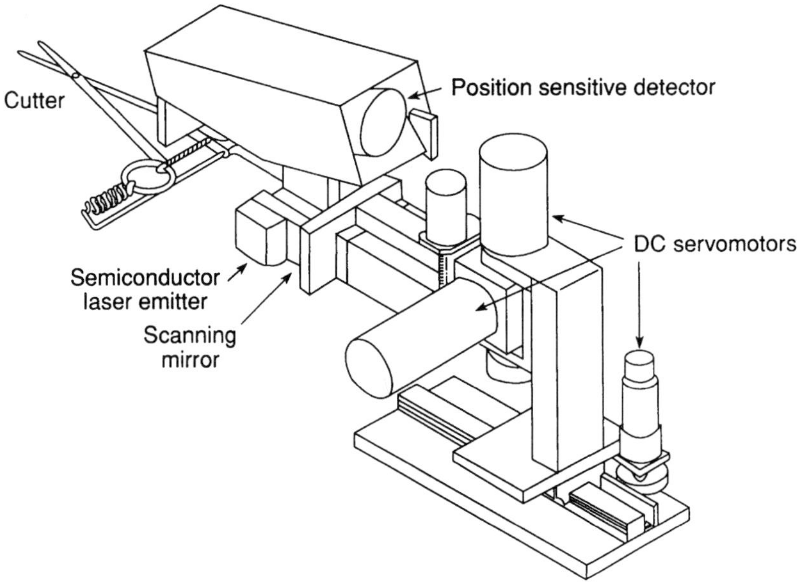

Fig. 3. Plantlet detection sensor (Watake and Kinase, 1988).

The upper two servomotors are used to change the direction of the scanning scope, and the lower one is used to traverse the sensor itself.

To detect the positions of plantlets a laser beam is used. A semiconductor laser emitter emits the laser beam, which is reflected by the scanning mirror upward and also reflected by the fixed mirror in the forward direction. The laser beam reflected by the plantlet is detected by a position-sensitive detector. The three-dimensional position where the laser beam is reflected by the plantlet is calculated.

Figure 4 explains the detection methods. In the actual system two mirrors including a scanning mirror are used, but in this figure they are simplified to one scanning mirror. From the laser emitter the beam is emitted and swung by the scanning mirror to a plantlet. If the beam is reflected by the plantlet, some of them are led to the PSD and the input beam angle Θ_2 is calculated. From Θ_2 and the scanning angle Θ_1, the position of the plantlet is easily calculated using the triangular method.

By measuring the stem in three positions along its length, it is also possible to detect the stem droops and bends. This information is used to determine the proper gripping positions and gripper attitudes.

A scissorlike cutter is attached at the tip of the plantlet detection sensor. It is driven through a wire rope from another part, and the cutter itself can be made smaller. When the position detective sensor is active, the cutter rotates not to interrupt the detection.

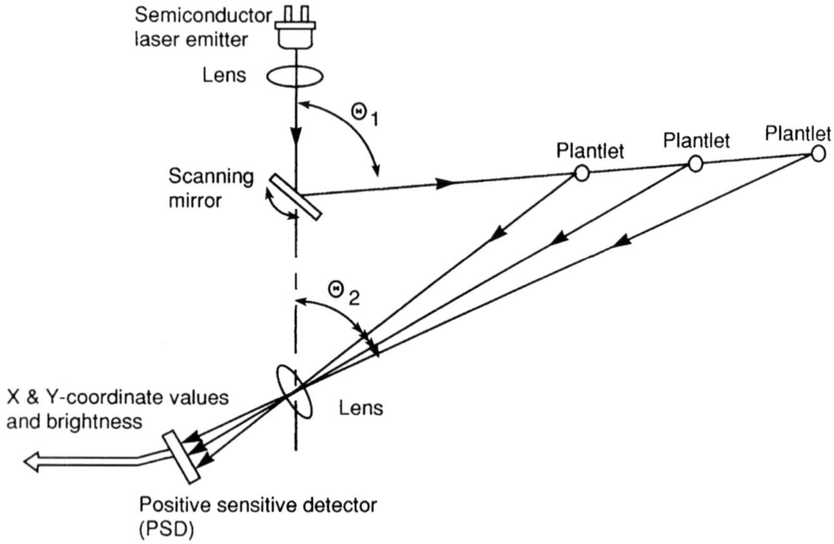

Fig. 4. Position detective method (Watake and Kinase, 1988).

D. Soft Gripper

Figure 5 shows the control method of the soft gripper. This tweezers-shaped gripper can handle very delicate objects such as plantlets or tofu without damaging them. The gripping force can be precisely controlled to the designated value. The strain-gauges pasted at the root of the fingers detect the strain-force which can be easily translated to gripping force. The DC servomotor is controlled in proportion to the difference between desired gripping force and measured gripping force.

When grasping plantlets, the gripping force can generally be set to 30 or 40 *g*. The actual value can be decided by experimentation.

Another important feature is that the gripper is controlled through wire ropes, so that the gripping part can be smaller and simpler.

Specification of the soft gripper is shown in Table I.

E. Robot Operation Procedure

1. First, the sensor checks the existence of the tray in which plantlet stems are grown.
2. The plantlet detection sensor detects the position of the first stem. The sensor measures the three positions of the stem along its length

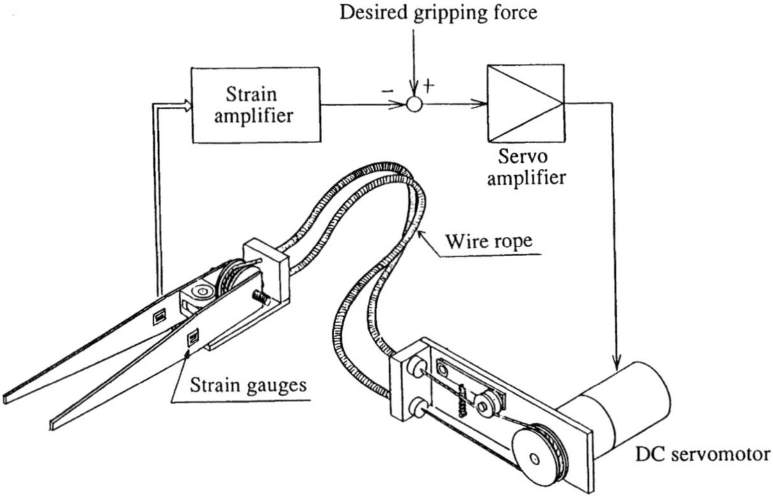

Fig. 5. Control method of soft gripper (Watake and Kinase, 1988).

by changing the direction of the sensor head, and then the direction and bend of the stem is modeled to an equation.

3. The cutting position of the first explant and the gripping attitude are calculated. The height of the cutting position is decided beforehand.
4. The robot approaches the position and the soft gripper grasps the stem. The cutter cuts off the stem.
5. The robot carries the cut-out stem to the settled position and transplants it in a new tray.
6. The system repeats these procedures to divide the stem into several portions (usually three portions), and finally it discards the remaining stem into the waste tray.
7. The robot system starts to detect the next plantlet to continue the procedure.

The required time for transplanting a cut-out stem including position detection and cutting is about 20 sec.

TABLE I
Specification of Soft Gripper

Control method	Gripping force feedback control
Driving method	Servomotor + wire + reducer
Gripping force	0–200 *g*
Accuracy	3 *g*
Response frequency	3 Hz

V. SECOND GENERATION ROBOT

A. Overview

This second generation robot was developed in March 1990 and was exhibited at the International Garden and Greenery Exposition held in Osaka, Japan.

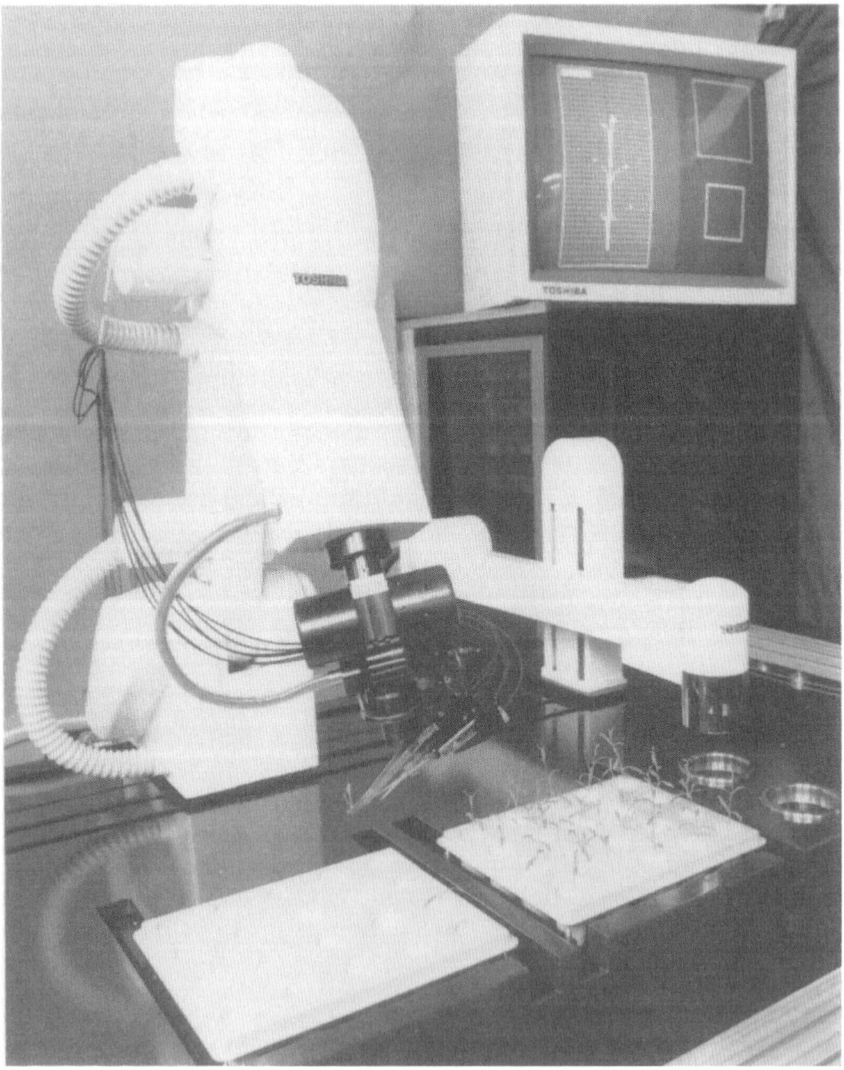

Fig. 6. A view of the second generation robot.

Figure 6 shows a view of this robot. The robot system consists of a sensing robot, a cutting robot, and surrounding apparatus. The sensing robot, which has a laser emitter, a scanning mirror, and a PSD, detects the shapes of plantlets on the culture tray. The other robot, which has a soft gripper and a scissorslike cutter, cuts off several portions from the plantlets and transplants them into the culture tray.

The technical improvements compared with the prototype robot are as follows:

1. The three-dimensional shape of each plantlet is measured, and the nodes of the plantlet are recognized. Then cutting positions are decided as shown in Fig. 7 so that each cut-off portion has a node.
2. The cutting robot and the sensing robot work concurrently unless there is no fear of collisions.
3. The required time for transplanting a cut-off stem including recognition has decreased to 15 sec, and in the near future it will be less than 10 sec.

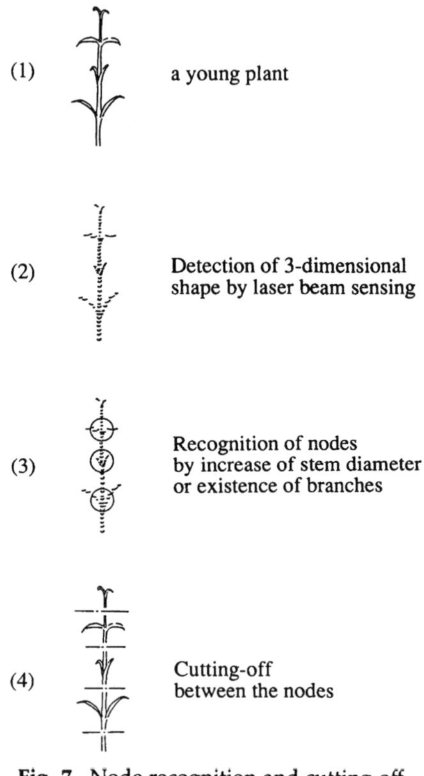

Fig. 7. Node recognition and cutting-off.

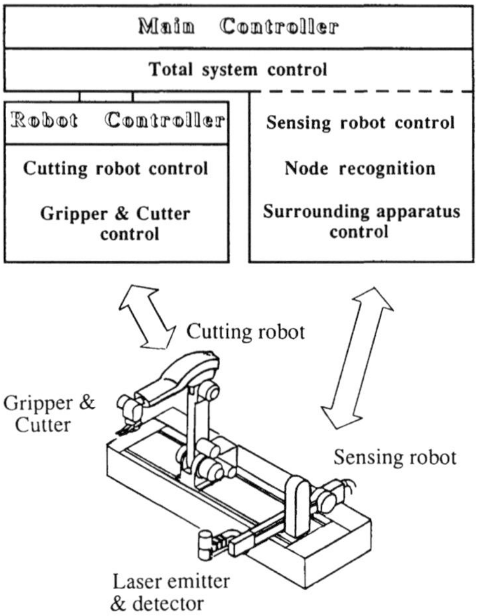

Fig. 8. Control structure of the second generation robot.

Figure 8 shows the control structure of the robot system. Both main controller and robot controller are composed of VME bus circuit boards and have Motorola's 32-bit microcomputers.

B. New Cutting Robot, Gripper, and Sensing Robot

This new robot, which has an excellent appearance, has been developed with the desirable capabilities of the prototype and has almost the same characteristics. The soft gripper and the cutter are also renewed.

For the sensing robot, a three-axes Cartesian coordinate robot has been developed. As each axis of this robot moves straight, not in rotation, the detective linearity is improved compared with the former model.

C. Three-Dimensional Shape Detection and Node Recognition

The sensing robot detects the three-dimensional shape of each plantlet through the following two steps, and from the shape the cutting positions are decided.

1. Coarse Detection

The emitted laser beam is swung almost horizontally by the scanning mirror to the assumed existence area of each plantlet (because of safety, the laser beam scanning surface is set 7 degrees below the level), and the reflected position data are gathered. The plantlet existence boundaries are previously assumed, and the data out of the boundaries are eliminated. Then the cross section of the plantlet along the scanned surface is detected as a line or lines shown in part (2) of Fig. 7. If the data have only one cross section line and the diameter is between the limits, then the cross section line is regarded as that of the main stem. The height of the scanning surface is lifted step by step from the root to the top and about 10 or 15 scans cover the length of each plantlet.

2. Fine Detection

At finer intervals the cross section forms are detected also from the root to the top of the plantlet. As the position of the main stem has been calculated by coarse detection, the laser scanning area is limited to the neighborhood of the main stem to save time.

3. Node Recognition

If stem diameters of some parts are larger than the decided value or some parts of the stem clearly branch, then those parts are assumed as possible nodes. And finally, according to some rules, such as the distance between nodes must be more than a value, the nodes to be cut off are confirmed, and the cutting positions are decided as shown in Color Plate 11.

D. Operation Procedure

1. The sensing robot measures the shape of a plantlet, recognizes the nodes, and decides where to cut.
2. The cutting robot grasps the stem and cuts it off.
3. The gripper carries the cut-out stem and transplants it on new culture medium.
4. The cutting robot continues to cut off the stem and transplant it until every node of the stem is cut off.
5. Concurrently, the sensing robot measures the next plantlet, avoiding the collision of both robots.

6. If all the plantlets on the cultured plantlet tray are cut off or the transplanted plantlet tray is filled with the cut-out stems, that tray is changed automatically. During the changing of the tray, the cutting robot washes and disinfects the cutter and the gripper.

VI. CONCLUSIONS

In the process of plantlet production through tissue culture, dividing cultured plantlets and transplanting them onto new culture media are necessary to increase their number. To automate these operations, a practical robot system has been developed. The robot system consists of a sensing robot, a cutting robot, and surrounding apparatus. The sensing robot detects the shape of each plantlet using the laser beam. The cutting robot divides each plantlet into several portions and transplants them onto the culture media. The important features of the system are:

1. the soft gripper which can control gripping force,
2. the sensing system which measures the three-dimensional shape of each plantlet, and
3. the methods of node recognition and cutting position decision.

ACKNOWLEDGMENTS

The authors wish to thank Dr. Kozai, Chiba University, and Dr. Takakuwa, Tokyo University, for their helpful support of this research.

REFERENCES

Kinase, A., and Tateishi, T. (1989). Development of Plantlet Micropropagation Robot "The Detection Method of Plantlet Node Position." *Proc. of 7th general assembly of the Robotic Society of Japan*. (*Japanese*)

Watake, H., and Kinase, A. (1988). Robot for Plant Tissue Culture, Robot No. 64, pp. 74–79. Japan Industrial Robot Association. (*Japanese*)

CHAPTER **13**

The Use of Automated Field Transplanters in Plant Propagation

Harold L. Brewer

Crop Systems Research Unit
U.S. Dept. of Agriculture
Coastal Plain Experiment Station
Tifton, Georgia

I. INTRODUCTION

Transplanting, the setting of young plants into fields rather than seeds, has been used for centuries (McKee, 1981). It is used in a variety of crops, such as trees, vegetables, rice, grasses, and sugar beets. Transplanting is used in every technologically advanced country and in most of the developing ones.

Transplanting accomplishes many things. It can overcome problems of establishing stands directly from seed, such as soil crusting, maintaining an environment conducive to seed germination, and the nonemergence of seedlings at uniform spacings. It is used to save resources, such as materials (seeds, fertilizers, pesticides, and water), field operations (seedbed preparation and cultivating), time (to accommodate short seasons or multi-cropping) and space (during early growth of the plants). Transplanting may also contribute to more timely and efficient harvesting.

Compared with distributing seeds in a field, planting seedlings is more labor- and management-intensive. It begins with the need to grow the seedlings, then to distribute them to fields, and finally to place the plants into conditions hospitable to their survival and continued growth. As with other agricultural operations, transplanting has undergone many transformations during this century. Gradually, operations have been mechanized, or at least labor has been aided in many of the tasks to be performed.

The advances in electronics and computers have made more realistic the visions that essentially all of the operations from seed or plant material to seedlings planted in fields can be automated, thus obtaining more uniform stands as well as reducing the human drudgery associated with the low-level, repetitive tasks involved. One area of increasing activity is the development of automatic transplanters. Most technologically advanced countries have some activities related to developing or adapting machines to drastically reduce the labor requirements of transplanting and to increase the quality of the planted stands in the field. Where these activities stand at this time and how they may eventually develop are the subjects of this chapter.

II. REVIEW

Periodically, reviews appear concerning transplanting (e.g., Boa, 1984a; Meeldijk, 1986; Kromer, 1986). None is comprehensive (i.e., contains all the material found in other reviews) nor can they be expected to present developments which occur after their individual publication dates. They are nevertheless valuable because they do summarize at least some activities probably not known to their readers and usually one or two new ideas can be gleaned from each. It is with these caveats that this review is written.

The functions of a field transplanter are related to the flow of seedlings from supply containers to planting mechanisms, then to the soil in which the seedlings are inserted. Seedlings arrive in close-packed, rectangular or circular arrays in a horizontal plane above the soil surface. They leave the transplanter in spaced-out linear arrays in the soil.

Most transplanters have humans riding them in the fields. The humans singulate seedlings from the supply containers, cull substandard seedlings, and place good seedlings in mechanical devices. The transplanter offers physical support for the humans, conveys seedlings from where they are obtained from the humans to the soil surface, opens the soil surface, places seedlings in the openings, and packs soil around them. The result is a high percentage of good seedlings planted in the field at rates up to about 60 seedlings per minute per row. Developers of automated transplanters have sought to accomplish the same results, usually at higher planting rates and with reduced labor. I will review some of these activities, although not all are considered because of space limitations or my not knowing about them.

Of the possible organizing schemes (historical, country of origin, crop, commercial development, or containers used), I will use containers because they are most useful, in my opinion, when comparing transplanters and when trying to project desirable paths to the future.

A. Same Container Used for Growing and Transplanting

The goal of most transplanter developers is to use the same containers for growing the seedlings, transporting them to the fields, and supplying them to the field transplanters. There are three major categories to consider.

1. Nonsegmented Containers

The mat used in Japanese machines to automatically transplant rice is an example. Seeds are sown in undifferentiated media and the resulting seedlings grown to transplant size. These mats are placed on transplanters which move through flooded paddies. Fingers pick off a small set of seedlings from the mat and push the set into the puddled soil (Shin-Norinsha Co., Ltd., 1971; Townsend and Tint, 1987). Thousands of units are in use, and they save considerable labor while not reducing crop yields or quality.

2. Segmented Containers—Fixed Cell

One of the most popular containers is that in which the cells are fixed with respect to one another. An example is the Speedling®[1] styrofoam

[1]Trade names are used in this publication solely for the purpose of providing specific information. Mention of a trade name, proprietary product, or specific equipment does not constitute a guarantee or warranty by the U.S. Department of Agriculture and does not imply approval of the named product to the exclusion of other products that may be suitable.

tray used extensively in the United States. Many others exist, but they are variations on the same theme—the cells in which individual seedlings are grown are unchangeably attached to one another in two-dimensional arrays.

Huang (1983) designed a special fixed-cell tray to use with the automated transplanter he developed during the 1960s and 1970s. Seedlings were removed through the bottom of the trays and dropped down a tube to the soil surface into which holes were made by a water jet.

Cayton *et al.* (1981), Rogers (1982), Olivar (1983), and de Groot (1986) reported the development of an automatic transplanter which used specially designed fixed-cell trays in which polymerized plugs were inserted and seedlings grown to the "just sprouted" stage. Several field machines were built and commercial plantings made with them. Only the plugs were handled as the seedlings were pushed out the top of the trays.

Another automatic transplanter which uses a specially designed fixed-cell tray is described by Rogers (1983), Branch (1986), and Klassen (1988). Several machines have been built which transplant commercially on a contract basis. Seedlings are pushed through the trays, out the bottom, directly into furrows.

One of the latest automated transplanters to be developed comes from Bud of California (de Groot, 1990), where several machines are transplanting commercially. Seedlings are delivered to the transplanters in specially designed fixed-cell trays. Seedlings are removed from the trays, out the top, and are delivered to the soil surface by grippers attached to a chain.

Lannen Plant Systems of Finland reports in its sales literature the development of an automatic transplanter which is a combination of a new automatic seedling feeder fitted to a transplanting mechanism the company had used previously on its hand-fed transplanters. Seedlings are removed from the top of specially designed fixed-cell trays and delivered to planting mechanisms.

Penley (1981) used a specially designed container that fits in the category "segmented containers—fixed cell" but with a variation. The seedlings are grown in modules (cakes) which are unsegmented at the bottom but are segmented in the uppermost layer of the cake so that seedlings can be grown in individual cells. These cakes are placed on the transplanter and automatically cut into rows from which individual seedling plugs are pulled for planting into furrows.

3. Segmented Containers—Bandoleer

A constantly recurring theme in containers is the bandoleer. A degree of freedom is gained over the fixed-cell arrays because cells are linked together in linear arrays. This feature makes it easy to imagine various ways to automatically feed the seedlings on transplanters.

Moden and Hauser (1982), Boa (1984b), and Hauser (1985) reported on transplanters which automatically fed bandoleers (in which the seedlings were grown) through planting mechanisms which removed individual seedlings from the bandoleers and placed them into the soil.

Masudo and Nambu (1979a, 1979b) obtained patents on a paper tube bandoleer transplanter, and Smith *et al.* (1984) reported on field trials using such a transplanter. The Williames Hi-Tech International Field Planter (Anonymous, 1987) uses specially designed bandoleers in which seedlings are grown to automatically feed seedlings to planting mechanisms.

Compton and Reid (1987) reported on an experimental transplanter which uses the bandoleer principle, thus showing the continued popularity of this container design.

B. Multiple Containers

Some researchers are working from the premise that more than one type of container may be desirable in the growing and transplanting of seedlings as long as the overall costs are competitive with single container–type systems. In one example of this (Eddington *et al.*, 1989), seedlings are grown in the soil, then bare-rooted and placed in supply tapes which can be fed through an experimental automatic transplanter. The two types of containers are unsegmented (soil beds) and bandoleer.

In another example (Brewer and Batal, 1983; Brewer, 1990), seedlings are grown in free cells which are supported in trays. When ready for transplanting, the seedlings are removed from the free cells and placed in shipping modules, which are used to transport the seedlings to the field and to feed them to planting devices on an experimental automatic transplanter. The two types of containers are free-cell and fixed-cell.

III. ANALYSIS

A. Evaluations

It is difficult to make evaluations and comparisons of differing transplanting systems for at least two reasons. First, evaluations are very similar to predictions. To judge a device, operation, or system is to make a statement as to whether people will find it useful in the future. The marketplace will determine when and if new transplanting systems are acceptable to growers.

Second, individuals and groups who have spent possibly years and hundreds of thousands of dollars on a particular approach are usually not pleased to have their efforts cavalierly dismissed as unacceptable. Again, the final arbiter is the marketplace, and I propose to leave it at that.

However, there are statements that can be made which may be useful to developers as they continue their activities.

B. Levels of Development

One way to classify efforts to automate transplanting is their level of development. Four general levels may be discerned in the work occurring worldwide.

1. Ideas

Generating ideas is the beginning of all developments and is an ongoing activity. Many times, an idea never gets implemented because it is judged to be defective or too expensive to try or it is eliminated for other reasons. Ideas about ways to automate transplanting are usually found, upon close examination, to rest upon underlying beliefs or principles, many times unstated, held by the person having the ideas. Lacking a common set of beliefs may explain why different people propose different solutions to supposedly the same problem—automating transplanting.

2. Experimental Models, Patents, and Papers

Initial implementation of an idea may result in a paper being written and possibly published, a patent being filed, and/or an experimental model being built. Examples of this level of development are Penley (1981), Boa (1984), Huang (1983), Hauser (1985), and Brewer (1990).

3. Field Trials

This stage is characterized by several machines operating in growers' fields on a commercial basis. Obviously, the machines work, but shortcomings are sometimes apparent, and corrective actions are in various planning stages. Some examples are those machines reported by Rogers

(1983) and Branch (1986), Olivar (1983) and de Groot (1986), and de Groot (1990). Transplanters at this level of development are usually manufactured by a single company and may not be available for purchase by the general public.

4. Commercially Successful Transplanters

Transplanters at this level of development are represented by hundreds or thousands of units operating extensively in growers' fields. Usually, several companies manufacture the units, and they are sold as widely as possible to the general public. An example is the Japanese rice transplanter reported by Shin-Norinsha Co., Ltd. (1971).

The normal progression of development is from ideas to commercially successful equipment. Of course, many developments at the first level never get to the second level, second level to the third, or the third level to the fourth. It is probably true that to get to any level, one must go through the levels preceding it. Attainment of a particular level does not guarantee permanence at that level nor advancement to the next. Even attainment of the commercially successful level is no guarantee of permanence because developments continue to occur.

C. Present State of Transplanter Industry

There are many "automated transplanters," and the developers of most of them believe that their system is as good as or better than competing systems. Most developers believe that, with a little more work, their system will play a (or the) major role in the future.

Millions of dollars and hundreds of worker-years have been spent developing automatic transplanters. Dozens of patents have been issued and are distributed among most of the technologically advanced countries. There is a continuing stream of inventions and new proposals concerning automated transplanting. At the present time, there are probably a dozen or more new ideas being tested.

Not one type of commercial transplanter has swept the field of hand-fed transplanters or completely dominated the planting of a significant crop where free markets exist.

Most automated transplanting systems are based on specialized components not readily adaptable to other systems. For example, the bandoleer transplanters cannot accept seedlings in fixed-cell trays, and vice versa. Most of the fixed-cell transplanters cannot interchange trays among themselves. There appear to be many individual efforts that show few signs of coalescing into a dominant form.

I conclude from these observations that today's transplanter industry is in about the same stage of development as the automobile industry was around 1900 (Flink, 1985).

IV. PROJECTIONS

A. Principles

The most certain statement I can make is that what we see today in automated transplanting is not all of what we will see in the future. Changes will occur and people will bring them about. These people will face a vast array of choices. Some way to select the more promising ones needs to be developed. One possibility is to enunciate principles which one believes will remain true for a considerable length of time. Principles help one to cut off whole lines of activity. This can be dangerous, however, because the "solution" may be left out of the search area. But the alternative is a random walk through an essentially infinite search space. Following are some principles which may be helpful.

1. Importance of Planting

Planting is a basic human activity, and continued innovations in its technology can lead to achievements only dimly glimpsed at this time (Brewer, 1986).

All advanced human societies have an absolute requirement for modern agriculture, and modern agriculture has an absolute requirement for planting. No planting, no agriculture. No agriculture, no advanced human societies. Innovations at such a basic level can have profound effects.

If this principle is true, then we need not necessarily build systems which interface with present ones. Newly designed transplanters do not have to fit currently available growing systems. Of course, new transplanters will require some forms of growing systems, but these may be modifications to present ones (just as highways were built to accommodate cars rather than cars being built to operate on horse trails or cow paths).

2. Cost

Something that is more costly initially is not automatically excluded from replacing something that is cheaper. Cars are more expensive than horses, yet they replaced them. The fact that bare-rooted plants are cheaper than containerized seedlings does not mean that bare-rooted seedlings will remain dominant.

Being higher up on the technology ladder usually costs more, not less. Thus, it costs more to ride a mile than to walk a mile. However, technology allows us to do things that we cannot do otherwise. We can walk 5 miles with little effort, but for us to ascend this far above the earth's surface requires technology.

3. Minimization

Give plants only the resources (e.g., light, space, nutrients) they require at a particular stage of their growth.

Transplanting is an application of this principle. A young plant doesn't require all the space in a field that it requires when mature. Therefore, space and time are conserved if seedlings are grown close-packed in greenhouses during their early stages of growth and then distributed to fields.

An alternative statement of the minimization principle is—cull out nonproductive material as early in the growth process as possible. This implies that we be able to sort material at any time, because we cannot predict which seeds will germinate or which newly emerged seedlings will grow into acceptable transplants. This statement of the principle makes it easy to see why it is undesirable to wait until plants get to the field transplanter to cull them. Nonproductive material is carried all the way through the growing phase up to the transplanter with consequent waste of resources and no economic return.

Application of this principle makes desirable the pregermination of seeds before they are given a cell and media to grow in. This is an example of a third statement of the minimization principle—make the plant material indicate that it needs more resources before committing any to it.

4. Billions of Seedlings

A transplanting system through its many units scattered across a dozen or more countries will handle billions of seedlings each year. If new transplanting systems result in significant increases in the use of transplanting, then the numbers could reach into the trillions. An individual producer could easily handle many millions of seedlings each year; therefore, it does not appear practical for humans to handle individual seedlings as a programmed part of production processes which transform seeds into seedlings planted in the field.

The minimum number of seedlings that a human should handle on a production basis is unknown at this time and is a matter of judgment. However, I will guess that individual humans should not handle seedlings in lots of less than 1000.

5. Automate

Seedlings need to be assessed and handled on an individual basis as well as in lots of less than 1000 (e.g., trays with 200 or so seedlings each). If humans are not to do it, then automated procedures are required.

For this to be practical, container systems need to be designed not only for growing quality seedlings but also for automated manipulation. The seedlings themselves need to be short and sturdy to facilitate their movement by machines (Brewer *et al.*, 1986). A single seed should be sown in each cell to ease the task of accepting or rejecting the resulting seedling by "machine intelligence." Once singulated (such as planting a single seed in each cell), plant materials should not be mixed again (such as pulling plants from a tray and placing them *en masse* in a shipping container).

6. Broadly Applicable Design

There are usually several ways to do something. For example, soil may be opened for seedling insertion by a knife (or shovel) making a continuous furrow, or a dibble punching holes at set intervals. This principle holds that if all other things are about equal, then design around the dibble. It can be used whenever the furrow opener is used, but not vice versa (e.g., a furrow opener would not work well in plastic covered beds or in many minimum-till situations).

Another example is the choice of raising a single seedling per cell or more than one. Design the system to raise single seedlings, because it can be used even where multiple seedlings are desired or acceptable, but not vice versa.

When designing the seedling singulation procedure, assume that tops will tangle if given a chance rather than designing a system in which seedlings are so short or smooth that the tops don't tangle.

Design automated systems for growing seedlings that will handle low germinating species. These will handle high germination seed lots as well, and thus the system will be more broadly applicable.

This is a nebulous principle, but its application can help developers avoid dead ends or severely limiting restrictions as their systems reach the field trial or commercial levels of development.

B. Possible Future Developments

1. What Will the Marketplace Want?

Generally, growers will want something which will help them deliver quality products to their customers, at good prices, and to do so reliably. To help them do this, they will want more economically transplanted fields of higher quality seedlings at 100% stands and precise spacings. Therefore, suppliers to the growers will strive to achieve certain performance standards in order to satisfy these demands:

1. use their resources economically without a lot of waste
2. use humans as supervisors vs. as laborers
3. seamless flow of materials from one part of the system to the next
4. transplanting rates above 200 seedlings/minute/row
5. set up initial plant stands to ease automating later operations, such as getting resources to plants (e.g., fertilizing, pest control, and harvesting)

2. What Resources Will Be Available?

If present trends continue, we can expect the following:

1. less costly and more sophisticated computers and electronic components
2. higher performance, but more costly, seeds and plant materials
3. more highly trained people who will expect to be well paid and permanently retained
4. better biological control materials and procedures
5. more advanced technology available, e.g., more sophisticated materials of construction
6. increasing surveillance of activities affecting everyone's environment

3. Self-Monitoring Transplanters

Present designs may lead to automated transplanters which plant almost perfect stands (99.99%) at rates exceeding 200 seedlings per row-minute. If they are to succeed, the transplanters will need to monitor their own operations for at least two reasons:

1. Machine elements will move too fast and there will be too many sequences to simultaneously monitor for humans to effectively observe, evaluate, and take action when errors or malfunctions occur.
2. The rare occurrence of errors will make the task of monitoring all

facets of the transplanting operation extremely boring to humans (even if they could respond quickly enough to be effective).

Self-monitoring by transplanters will include determining whether a seedling has been transplanted at each desired site in the field, whether seedlings are successfully moving between components on the transplanter, and whether performance is starting to deteriorate at some particular point (e.g., because of buildup of trash on some surface). The control unit on the automated transplanter will receive inputs, evaluate the inputs, and take action accordingly. Actions might be to continue as before, take corrective action and continue, or shut down the operation.

Electronic components and computers are required for this monitoring activity; therefore, future automated transplanters can be expected to have these technologies on board.

4. Autonomous Field Vehicles

Planting at high rates requires containers holding seedlings to cycle through the transplanter quickly. Self-monitoring by the transplanter and automated feeding of seedlings require fewer humans in the field. Therefore, mechanized ways of transferring seedlings from the highway vehicles (which deliver seedlings from greenhouses to fields) to the automated transplanters and then back again must be devised.

The high rate of transplanting will also make it difficult for enough seedlings to be loaded onto the transplanter at one time for it to operate for more than a few minutes without replenishment of the seedling supply.

Mechanized loading of seedling containers onto the transplanter and the need to reload at several-minute intervals lead to the possibility of two autonomous vehicles making up the field transplanting system. The first vehicle will be the transplanter itself which moves through the field following an internal map of the field, carrying out a set of stored instructions as to the job to be done and self-monitoring to determine the correctness of its actions. The supply of seedlings will be enough to keep the transplanter going less than a minute. No human need be aboard.

The second vehicle, a supply shuttle, will move full seedling containers from the highway vehicle to the transplanter, act as a source of seedlings (for up to 5 minutes) for the transplanter, and move empty seedling containers back to the highway vehicle. Containers are automatically loaded onto the shuttle at the highway vehicle. The shuttle moves to the transplanter wherever it is in the field, locks onto it so that they move as a unit, supplies full seedling containers to the transplanter and removes empty containers, unlocks from the transplanter, and returns to the highway vehicle for unloading. The shuttles may also supply other materials to the transplanter, such as water or nutrient solution to be added to each seedling site. A human supervisor may be positioned at the highway vehicle, receiving reports from the autonomous vehicles and issuing commands to them via high-level language interfaces (e.g., English, Japanese). At least two shuttles will be required per transplanter.

5. Autonomous Transplanters as "Black Holes"

If the transplanting system is divided into two vehicles, one of them supplying the other, then relatively small units can be designed. Small units can allow increased speed and maneuverability in the field. One-row transplanters will not have enough transplanting capacity and will not effectively use the necessary computer/electronic components to be economically justified. Greater than four-row transplanters introduce maneuvering problems in the field. Therefore, two-, three-, or four-row transplanters will be the most common because they can utilize the computer/electronics components efficiently and have sufficient transplanting capacity to be of commercial interest.

A further advantage of having, for example, two four-row transplanters vs. one eight-row transplanter is that if some malfunction incapacitates the transplanter, then with a single eight-row unit, all transplanting stops, whereas one of the four-row transplanters could keep going.

The transplanters being autonomous leads to essentially no labor input, which releases the machines to work 24 hours a day if weather and field conditions permit. A four-row transplanter supported by supply shuttles working around the clock should be able to transplant hundreds of thousands of seedlings per day at 99.99% success rates. Such transplanters will appear to be "black holes" to the facilities growing the seedlings for the field. Therefore, seedlings will be grown reasonably close to where they will eventually be transplanted. The effects of minimization and automation may make this more practical than it has been in the past.

V. CONCLUSIONS

There is much activity in the attempts to automate the transplanting of seedlings in fields. The past two decades have seen the actualization of what were previously merely dreams. I believe it reasonably safe to predict that developmental activities will continue until the major portion of all transplanting will be automated. Which developments will be successful during the first decade of the next century is unknown at this time, but some of the possibilities are:

1. None of the present developments will survive into the next century.
2. One or two of the present transplanters in various stages of development will survive and become dominant.
3. Many different transplanter configurations will coexist.
4. New designs will come along that will make obsolete that which has been accomplished to date.

As a result of continued activities in automating processes and in other areas of plant propagation, the industry will be transformed from that which we know today.

REFERENCES

Anonymous. (1987). Automatic transplanter makes California debut. *American Vegetable Grower* (May), **78.**

Boa, W. (1984a). The development of transplanting systems. *In* "Crops Establishment: Biological Requirements and Engineering Solutions." (M.K.V. Carr, ed.) Assoc. of Applied Biologists, Nat. Veg. Research Station, Wellesbourne, United Kingdom.

Boa, W. (1984b). The design and performance of an automatic transplanter for field vegetables. *J. Agric. Engin. Res.* **30,** 123–130.

Branch, G. (1986). A commercially successful automatic transplanter. ASAE Paper No. 86-1020, ASAE, St. Joseph, Michigan 49085.

Brewer, H. L. (1986). Automated seedling production. *Acta Hort.* **187,** 53–59.

Brewer, H. L. (1990). Experimental static-cassette automatic seedling transplanter. ASAE Paper No. 901033, ASAE, St. Joseph, Michigan 49085.

Brewer, H. L., and Batal, K. M. (1983). A free-cell system of raising seedlings to conserve greenhouse energy use. *Energy Agric.* **2,** 137–151.

Brewer, H. L., Batal, K. M., Jones, P., Jones, J. W., and Allen, Jr., L. H. (1986). Raising short sturdy seedlings to facilitate automated handling. ASAE Paper No. 86-4061, ASAE, St. Joseph, Michigan 49085.

Cayton, D. W., Armstrong, E. C., and MacKenzie, A. B. (1981). Dibble tube soil plug planter. U.S. Patent No. 4,294,179.

Compton, B. A., and Reid, W. S. (1987). An automatic media-block transplanting system. Contribution No. I-924, Engineering and Statistical Research Centre, Research Branch, Agriculture Canada, Ottawa K1A OC6.

de Groot, P. F. (1986). Recent innovations in the stabilized soil plug automatic transplanter. ASAE Paper No. 86-1019, ASAE, St. Joseph, Michigan 49085.

de Groot, P. F. (1990). The "Centipede" automatic transplanter. ASAE Paper No. 901035, ASAE, St. Joseph, Michigan 49085.

Eddington, D. L., Suggs, C. W., and Seabock, T. R. (1989). An automatic bare root transplanter. ASAE Paper No. 89-1081, ASAE, St. Joseph, Michigan 49085.

Flink, J. J. (1985). Innovation in automotive technology. *Am. Sci.* **73,** 151–161.
Hauser, V. L. (1985). An automatic grass transplanting machine. Transactions of the ASAE **28(6),** 1777–1782.
Huang, B. K. (1983). Systems engineering in precision automatic transplanting. *Agric. Mechan. in Asia, Africa and Latin America* **14(1),** 11–19.
Klassen, P. (1988). Full-service transplanting. *American Vegetable Grower* **36(4),** 8–9.
Kromer, K. H. (1986). Transplanting systems. *Acta Hort.* **187,** 81–86.
Masuda, A., and Nambu, T. (1979a). Continuous paper cylinder assembly and method of separating the same and transplanting apparatus for the same. U.S. Patent 4,132,337.
Masuda, A., and Nambu, T. (1979b). Apparatus for separation and transplanting paper tube seedlings from a continuous paper tube seedling assembly. U.S. Patent No. 4,167,911.
McKee, J. M. T. (1981). Physiological aspects of transplanting vegetables and other crops. I. Factors which influence re-establishment. *Hortic. Abst.* **51,** 265–272.
Meeldijk, B. P. (1986). Plantmachines voor kluitplanten. Landbouwmechanisatie **37(4),** 386–389.
Moden, Jr., W. L., and Hauser, V. L. (1982). An automatically fed bandoleer transplanter. *Trans. ASAE* **25(4),** 864–867.
Olivar, M. A. (1983). Mechanical transplanting. *The Grower* (May), 26–28.
Penley, P. A. (1981). High-speed transplanter. U.S. Patent No. 4,289,080.
Rogers, H. T. (1982). Plugging away at labor-less transplanting. *American Vegetable Grower* (May), 8–9.
Rogers, H. T. (1983). Mechanical transplanting system begins commercial operation. *American Vegetable Grower* (Nov), 26–29.
Shin-Norinsha Co., Ltd. (1971). The latest mechanization of rice transplanting in Japan. *Agric. Mechanization Asia* **2,** 119–133, Autumn.
Smith, J. A., Yonts, C. D., Wilson, R. G., Kerr, E. D., and Robb, J. G. (1984). Paperpot system for mechanical sugarbeet transplanting. ASAE Paper No. 84-1516, ASAE, St. Joseph, Michigan 49085.
Townsend, J. S., and Tint, S. (1987). Design improvements for a rice transplanter indexing mechanism. ASAE Paper No. 87-5052, ASAE, St. Joseph, Michigan 49085.

Index